Lessons from Others for Future U.S. Army Operations in and Through the Information Environment

Christopher Paul, Colin P. Clarke, Michael Schwille, Jakub P. Hlávka, Michael A. Brown, Steven S. Davenport, Isaac R. Porche III, Joel Harding

Prepared for the United States Army
Approved for public release; distribution unlimited

For more information on this publication, visit www.rand.org/t/RR1925z1

Library of Congress Cataloging-in-Publication Data is available for this publication.
ISBN: 978-0-8330-9815-3

Published by the RAND Corporation, Santa Monica, Calif.

Cover photos (clockwise from top left): Giorgio Montersino via Flickr (CC BY-SA 2.0);
U.S. Air Force photo by Airman 1st Class Adawn Kelsey;
U.S. Air Force photo by Tech Sgt John Gordinier;
U.S. Air National Guard photo by Master Sgt Andrew J. Moseley;
Russian Ministry of Defence (CC BY 4.0);
North Korean national media

Preface

The information environment (IE) has become more complicated, more extensive, more ubiquitous, and more important to the outcomes of military operations than ever before. Because the U.S. Army does not have the same level of overmatch in the IE that it maintains in the land domain, there is room for improvement and there are opportunities to learn from others. This project sought to draw lessons from the efforts of others (both nation-state and nonstate actors) in and through the IE for future U.S. Army force planning and investment. These case studies of other forces can help the Army (1) identify capabilities and practices that it should consider adopting and (2) identify adversary capabilities and practices that it must be prepared to counter in future operations in and through the IE.

The study drew lessons from the practices and principles for operating in the IE in 12 cases: four allies, four state actors of concern, and four nonstate actors of concern. Eight practices and principles are worth emulating:

- generous resourcing of information power and information-related capabilities (IRCs)
- giving prominence to information effects in operational planning and execution
- increasing the prestige and regard of IRC personnel
- integrating physical and information power
- extensive use of information power in operations below the threshold of conflict
- employing the concept of getting the target to unwittingly choose one's own preferred course of action
- carefully recording and documenting one's own operations
- high production values.

There appear to be three reasons for gaps between U.S. Army capabilities and the best-in-breed capabilities of others: capacity gaps, conceptual gaps (where effective concepts have not been adopted), and gaps stemming from authority or ethical constraints.

The study makes the following recommendations to the Army:

- Give effects in and through the IE greater emphasis and priority.

- Promote a view of information power as part of combined arms.
- Routinize and standardize the processes associated with operations in and through the IE to be consistent with and part of other routine staff processes.
- Tie political, physical, and cognitive objectives together coherently in plans, and communicate those compound objectives clearly to maneuver forces.
- In coordination with the U.S. Department of Defense and the broader U.S. government, seek expanded authorities to operate in the IE short of declared hostilities.
- Bring more information operations (IO), military information support operations, and other IRCs out of the reserves.
- Tear down or move the firewall between public affairs and other IRCs.
- Increase the volume and efficacy of education and training in operations in and through the IE and information power to reflect the increasing importance of these capabilities for the U.S. military and the nation, as well as the greater role they will play in future conflicts.
- Take steps to close capacity gaps in key capability areas—including cyber, influence, operations security, and military deception—by making IO and IRC career fields and military occupation specialties larger, more attractive, and more prestigious.

This report should be of interest to Headquarters, U.S. Department of the Army, personnel who are responsible for force planning and capability investment for cyber operations, information operations, and IRCs, as well as information operations officers and members of both planning and operations staffs who might be called to operate alongside or against any of the forces studied. A companion report, *Lessons from Others for Future U.S. Army Operations in and Through the Information Environment: Case Studies*, presents the full case studies and detailed lessons for U.S. Army operations.

This research was reviewed and approved by RAND's Institutional Review Board (the Human Subjects Protection Committee). RAND operates under a "Federal-Wide Assurance" (FWA00003425) and complies with the Code of Federal Regulations for the Protection of Human Subjects Under United States Law (45 CFR 46), also known as "the Common Rule," as well as with the implementation guidance set forth in DoD Instruction 3216.02.

This research was sponsored by Deputy Chief of Staff, G-3/5/7, and conducted within the RAND Arroyo Center's Strategy, Doctrine, and Resources Program. RAND Arroyo Center, part of the RAND Corporation, is a federally funded research and development center sponsored by the United States Army.

The Project Unique Identification Code (PUIC) for the project that produced this document is RAN167283.

Contents

Preface ... iii
Tables ... vii
Summary ... ix
Acknowledgments ... xv
Abbreviations ... xvii

CHAPTER ONE
Introduction ... 1
Challenges Facing Operations in and Through the Information Environment ... 1
Opportunities for the Army ... 3
Study Approach ... 3
Key Terms and Their Use ... 5
How This Report Is Organized ... 7

CHAPTER TWO
Summary and Overview of the Cases ... 9
Topics Covered in the Case Studies ... 9
Israel ... 10
NATO ... 10
Canada ... 11
Germany ... 12
China ... 12
North Korea ... 13
Iran ... 14
Russia ... 14
Hezbollah ... 15
Al-Qaeda ... 15
ISIL/Daesh ... 16
Mexican Drug-Trafficking Organizations ... 17

CHAPTER THREE
Comparative Analysis: Common Themes in the Cases 19
Capability Areas in Which Others Excel 19
Common Concepts and Principles Across the Cases 24
Key Takeaways 29

CHAPTER FOUR
Comparative Analysis: Distinctive Features of the Cases 31
Israel 31
NATO 32
Canada 33
Germany 34
China 35
North Korea 37
Iran 37
Russia 38
Hezbollah 39
Al-Qaeda 39
ISIL/Daesh 40
Mexican Drug-Trafficking Organizations 40

CHAPTER FIVE
Conclusions and Recommendations 41
Recommendations 42

References 45

Tables

S.1. Capability Areas in Which Others Have Excelled.................................. xi
S.2. Common Concepts and Principles Governing Others' Operations in the IE.... xii
3.1. Capability Areas in Which Others Have Excelled.................................. 20
3.2. Common Concepts and Principles Governing Others' Operations in the IE.... 25

Summary

Challenges and Opportunities in the Information Environment

Over the past few decades, the global information environment (IE) has seen significant changes driven by use patterns and evolving technology, including the merging of the wired and wireless worlds and the continuous expansion of the cyberspace domain. The contemporary IE can be characterized by its unprecedented breadth, depth, and complexity, but also by its ubiquity, hyperconnectivity, and exponential growth. Of special significance is the emergence of developing countries in the IE, their participation boosted by the proliferation of wireless technologies.

Even as the IE becomes more complex and extensive, popular perceptions, attitudes, and behaviors shaped in and through it continue to grow in strategic importance. People, governments, troops, and leaders are connected in ways that were unimaginable two decades ago. The accord of legitimacy to a government or military force's actions is now much more immediate and consequential. No longer can military operations focus strictly on desired physical outcomes.

Competitors, adversaries, and potential adversaries understand these changes and are becoming more active in and through the IE and more adept at achieving operational and strategic objectives. The diffusion of information technology has not only increased the range and scope of effects available in and through the IE, but it has also lowered the barriers to entry. Less sophisticated state actors and even nonstate actors have acquired capabilities that were previously available only at great expense to the most advanced nations.

The very challenges inherent in the IE create an opportunity for the U.S. Army—namely, the chance to learn from the experiences of others in contending with new developments in this dynamic environment. This report presents lessons for future U.S. Army organization and capabilities as it prepares for operations in and through the IE. These lessons are drawn from 12 case studies of others' efforts in this area, summarized in Chapter Two and discussed in depth in a companion report, *Lessons from*

Others for Future U.S. Army Operations in and Through the Information Environment: Case Studies.[1]

Study Approach

To help identify investment and development targets for U.S. Army information operations and information-related capabilities (IRCs), we asked three strategic questions:

1. What practices or capabilities have U.S. allies employed effectively in the IE, and which of these could the Army adopt?
2. What information-related practices or capabilities have U.S. adversaries or potential U.S. adversaries used effectively, and which of these could the Army adopt?
3. What are adversaries or potential adversaries doing in this space that the Army cannot consider doing because of ethical or legal constraints but that it must be ready to counter nonetheless?

We answered these three questions through a careful examination of the doctrine and practices of selected allies, adversaries, and potential adversaries. The case studies focus on four allies (Israel, the North Atlantic Treaty Organization [NATO], Canada, and Germany), four state actors of concern (China, North Korea, Iran, and Russia), and four nonstate actors of concern (Hezbollah, al-Qaeda, the Islamic State of Iraq and the Levant [ISIL]/Daesh, and drug-trafficking organizations [DTOs] in Mexico).

Findings

Our analyses revealed that the actors in several of the cases excelled in multiple capability areas. In fact, a cursory examination of Table S.1 reveals that several cases featured a noteworthy capability in every aspect of operations in the IE. For example, both China and Russia earned a check or ½ in every column. Among the nonstate actors, Hezbollah and ISIL were noteworthy, having excellent capabilities in more than half of the listed capability areas. Among U.S. allies, Germany was a standout, though it had fewer capability areas of excellence than almost all the actors of concern. Different cases demonstrated different areas of excellence, but all cases had at least some.

Through further analyses, we considered the practices and principles employed across the cases, with the intent of matching patterns of practice with observed capability strength and operational success. Table S.2 depicts common concepts and principles observed in the cases studied.

---

[1] Christopher Paul, Colin P. Clarke, Michael Schwille, Jakub P. Hlavka, Michael A. Brown, Steven S. Davenport, Isaac R. Porche III, and Joel Harding, *Lessons from Others for Future U.S. Army Operations in and Through the Information Environment: Case Studies*, Santa Monica, Calif.: RAND Corporation, RR-1925/2-A, 2018.

Table S.1
Capability Areas in Which Others Have Excelled

Case	Excels in press relations/media operations/public relations/government broadcasting	Excels in influence	Excels in leveraging narrative	Excels in deception/stratagem/manipulation	Excels in electromagnetic spectrum operations/electronic warfare/jamming	Excels in cyberoperations	Excels in OPSEC/secrecy/denial/information security	Excels in censorship/information control	Excels in the use of maneuver and fires as IRCs	Excels in outsourcing/use of proxies or militias/franchising for IRCs	Excels in leveraging social media/new media/now media
Allies											
Israel					½	√					
NATO							½				
Canada							½				
Germany	√				½		√				√
State actors of concern											
China	√	√	½	√	√	√	√	√	√	√	√
North Korea			√			√	√	√			
Iran						√	½	½		√	
Russia	√	√	√	√	√	√	½	√	√	√	√
Nonstate actors											
Hezbollah	√	√	√				½		√		√
Al-Qaeda							½		√		√
ISIL/Daesh	√	√	√				½	½	√	√	√
Mexican DTOs		√	√					½	√		

KEY:
Blank = not doing/not significantly present/not excellent compared with other cases
√ = significantly present in or characteristic of case
½ = partially present in or somewhat characteristic of case

NOTE: OPSEC = operations security.

Table S.2
Common Concepts and Principles Governing Others' Operations in the IE

Case	Information power capabilities generously resourced	Information effects given priority/prominence in operations	Information power–related personnel accorded prestige, otherwise well rewarded	IRCs collected together in single organization or command	Integration of informational and physical power	Extensive use of informational combat power in operations below the threshold of conflict	Employed concept of getting target to unwittingly choose preferred course of action	Significant focus of information efforts on own domestic/internal audience	Careful recording/documenting of own operations	Complete commitment to "white" information—no influence or manipulation	No compunction about falsehood or manipulation	High production values for video, digital, and other products
Allies												
Israel								√	√			
NATO								½		√		½
Canada				½								
Germany	½	½		½	½			½	√	√		√
State actors of concern												
China	√	√		√	√	√	√	√			√	√
North Korea	√				√	√		√			√	
Iran	½	½			½	√		√			√	
Russia	√	√			√	√	√	√			√	√
Nonstate actors												
Hezbollah	√	√	√		√	√		√	√		√	√
Al-Qaeda	½	√			√	√					√	½
ISIL/Daesh	√	½	√	½	√	√		√			√	√
Mexican DTOs		√			√			√			√	

KEY:
Blank = not doing/not significantly present/not excellent compared with other cases
√ = significantly present in or characteristic of case
½ = partially present in or somewhat characteristic of case

Our results revealed several practices and principles worth emulating, as well as several more practices and principles that are sinister and should not be emulated but that the Army must be prepared for. Practices and principles worth emulating include the following:

- generously resourcing information power and IRCs
- giving prominence to information effects in operational planning and execution
- increasing the prestige IRC personnel
- integrating physical and information power
- using information power extensively in operations below the threshold of conflict
- employing the concept of getting the target to unwittingly choose one's preferred course of action
- carefully recording and documenting one's own operations
- producing high-quality video, digital, and other information products.

Actors of concern have a number of advantages in the IE. One is that they are unconstrained by ethics, law, or policy. Practices and principles that are unacceptable under American laws and values but common and effective—and that actors of concern will continue to leverage—include the following:

- a focus on maintaining the support of domestic audiences by any means necessary
- a lack of compunction about falsehood or manipulation
- censorship and information control.

Conclusions and Recommendations

Our findings led us to conclude that the U.S. Army should adopt as many of the effective practices as possible, eschewing only those that are inconsistent with American laws or values.

Furthermore, where there are gaps between U.S. Army capabilities and the best-in-breed capabilities observed among others, there appear to be three reasons. First, some capability gaps are actually *capacity* gaps: Technical capability or capability proficiency is similar between the Army and other actors, but others have invested more and thus enjoy greater quantities or higher levels of the relevant capability. Second, some capability gaps are actually *conceptual* gaps: Others have adopted effective principles and practices that the Army has not yet adopted. Third, some capability gaps are due to differences in constraints: Some other actors operate in a virtually unconstrained fashion, while the Army operates in accordance with its authority and guided by statutes, policies, and ethics. Obviously, the first two reasons for capability gaps are more amenable to gap closure; gaps stemming from others' use of illegal or unethical practices must be bridged in other ways.

The results of this study point to the following recommendations for the Army:

- Give effects in and through the IE greater emphasis and priority.
- Promote a view of operations in and through the IE as part of combined arms.
- Routinize and standardize the processes associated with the IE to be consistent with and part of other routine staff processes.
- Tie political, physical, and cognitive objectives together coherently in plans, and communicate those compound objectives clearly to maneuver forces.
- In coordination with the U.S. Department of Defense and the U.S. government more broadly, seek expanded authorities to operate in the IE short of declared hostilities.
- Bring more information operations (IO), military information support operations, and other IRCs out of the reserves.
- Tear down or move the firewall between public affairs and the other IRCs.
- Increase the volume and efficacy of education and training in operations in and through the IE and information power to reflect the increasing importance of these capabilities for the U.S. military and the nation, as well as the greater role they will play in future conflicts.
- Take steps to close capacity gaps in key capability areas, including cyber, influence, operations security, and military deception, by making IO and IRC career fields and military occupational specialties larger, more attractive, and more prestigious.

Acknowledgments

We would like to express our gratitude to our various points of contact and interlocutors in the sponsoring office, originally Headquarters, U.S. Department of the Army, Department of the Army Military Operations, Cyberspace and Information Operations Division, including COL Carmine Cicalese, COL Jeffery Church, LTC Joel Humphries, Patrick Scribner, and LTC John Zollinger. We also thank those whose thoughts and input helped us better understand Army capabilities for operations in and through the IE, including LTC Matt Tieszen, CPT Peter Graham, COL John Bircher, LTC Pam Tindal, LTC David Ambrose, Richard Badley, LTC Joseph Hammond, CPT Scott Fischer, and Robert Hill. Our thanks, too, to our quality assurance reviewers, Edward Fisher at the IO Center for Research at the Naval Postgraduate School and Arturo Munoz at RAND, whose valuable comments improved the final version of this report. We owe a debt to Maria Falvo for her support and assistance with the report's citations and bibliography. Finally, we thank the RAND editing and publications team, including Todd Duft and Lauren Skrabala, for their tireless efforts to bring this report to its clean and much more readable final form.

Abbreviations

CAF	Canadian Armed Forces
DoD	U.S. Department of Defense
DTO	drug-trafficking organizations
EW	electronic warfare
IDF	Israel Defense Forces
IE	information environment
IO	information operations
IRC	information-related capability
ISIL	Islamic State of Iraq and the Levant
MILDEC	military deception
MISO	military information support operations
NATO	North Atlantic Treaty Organization
OPSEC	operations security
PA	public affairs
PLA	People's Liberation Army
PSYOP	psychological operations

CHAPTER ONE

Introduction

This report presents lessons for future U.S. Army organization and capabilities for operations in and through the information environment (IE) drawn from 12 case studies of others' efforts in this area. This chapter reviews the challenges posed by these types of operations, the opportunities for the Army inherent in some of those challenges, the research approach used for this study, and relevant terminology for interpreting the case studies.

Challenges Facing Operations in and Through the Information Environment

Over the past few decades, the IE has seen significant changes driven by use patterns and evolving technology, including the merging of the wired and wireless worlds and the continued expansion of the cyberspace portion of the IE.[1] The contemporary IE can be characterized by its unprecedented breadth, depth, and complexity, but also by its ubiquity, hyperconnectivity, and exponential growth.[2] Emerging technologies, including the Internet of Things, artificial intelligence, and big data, promise still greater changes.[3]

---

1 Isaac R. Porche III, Christopher Paul, Michael York, Chad C. Serena, Jerry M. Sollinger, Elliot Axelband, Endy M. Daehner, and Bruce J. Held, *Redefining Information Warfare Boundaries for an Army in a Wireless World*, Santa Monica, Calif.: RAND Corporation, MG-1113-A, 2013.

2 U.S. Department of Defense (DoD), *Department of Defense Strategy for Operations in the Information Environment*, Washington, D.C., June 2016, p. 4; Isaac R. Porche III, Bradley Wilson, Erin-Elizabeth Johnson, Shane Tierney, and Evan Saltzman, *Data Flood: Helping the Navy Address the Rising Tide of Sensor Information*, Santa Monica, Calif.: RAND Corporation, RR-315-NAVY, 2014.

3 *Big data* refers to large, usually interconnected, data sets used for computational analyses of patterns, trends, and associations that are more difficult to find in single data sets. The *Internet of Things* is a contemporary concept in which physical objects have network connectivity, allowing them to send and receive data. Common and existing examples include thermostats that can be adjusted remotely and refrigerators that report when contents have exceeded their use-by date (and perhaps add that item to a shopping list or even automatically order a replacement), but future applications promise a much greater scope.

The world is moving rapidly toward levels of connectivity that will further change how and where people associate and gather, share, and consume information. The trend toward true ubiquity means that the IE will touch almost everyone, everywhere. This matters not only for the ability to pass information among nodes in the IE but also for sensing and recording at these nodes. As the U.S. Department of Defense strategy document on operations in the IE noted, "[T]he ubiquity of personal communications devices with cameras and full-motion video allows much of the world to observe unfolding events in real time. These same capabilities can be utilized by adversaries for operational purposes, as well as for propaganda and disinformation."[4]

As the IE becomes more complex and extensive, popular perceptions, attitudes, and behaviors shaped in and through it continue to grow in strategic importance. Today, publics, governments, troops, and leaders are connected in ways that were unimaginable two decades ago. The accord of legitimacy for a government or military force's actions is now much more immediate and consequential. Military operations can no longer focus strictly on desired physical outcomes. Carl von Clausewitz's assertion that war is politics by other means continues to be true, and the contemporary IE makes sure that political actions now play out much more rapidly, and with many more participants, than in the past. Indeed, war is still politics by other means, and politics is inextricably linked with the IE.

Competitors, adversaries, and potential adversaries understand these changes and are becoming more active in and through the IE and more adept at achieving related operational and strategic objectives. The diffusion of information technology has not only increased the range and scope of effects available in and through the IE, but it has also lowered the barriers to entry. Less sophisticated state actors and even nonstate actors have acquired capabilities that were previously available only at great expense to the most advanced nations.

Violent extremist organizations gather recruits and financial support through the Internet and social media, and their propaganda intimidates, undermines, and weakens their foes. Russia's operations in Ukraine and elsewhere have demonstrated its commitment to a new mode of competition, following incrementalist or gradualist strategies and including a wide range of operations that aim for effects in not only the physical but also the perceptual, moral, and mental realms. Chinese doctrine and emerging practice call for preparation for not just physical modes of war but also legal warfare, warfare within public opinion, and psychological warfare. These actors "disseminate truthful, biased, and false information using digital technologies and access to global audiences to recruit, to gain support and sustainment, and to exploit, disrupt, and delegitimize U.S. and coalition operations."[5]

---

[4] U.S. Department of Defense, 2016, p. 4.

[5] U.S. Department of Defense, 2016, p. 2.

Because of the expanding scope and complexity of the IE, the reduced entry cost of gaining asymmetric capabilities to affect it, and the host of actors exploiting the IE to gain the advantage, the Army does not have the same level of definitive overmatch in the IE that it maintains in the land domain. Moreover, despite some tactical successes in Afghanistan, the U.S. military writ large still struggles with the integration and harmonization of much information operations (IO) doctrine while also grappling with serious challenges to measuring the effectiveness of its operations.[6]

Opportunities for the Army

The fact that the United States cannot claim the best-in-breed title in all aspects of operations in and through the IE presents a tremendous opportunity that is unavailable in areas where the United States is unmatched: The U.S. Army can learn from others' successes. The postures and actions of U.S. partners, adversaries, and potential adversaries in and through the IE have exposed the Army to new concepts, capabilities, and practices. By scrutinizing those actors and their actions, the Army can see how others organize and prepare to fight and gain advantage in the IE. In this report and accompanying case studies, we identify successful concepts, approaches, and capabilities already used by others and provide recommendations for their incorporation and for the Army investments to promote the effectiveness of future operations.

Study Approach

To help guide future investment and development planning for U.S. Army capabilities for operating in and through the IE (including IO and information-related capabilities [IRCs]), this study asked three questions:

1. What practices or capabilities are being employed effectively in the IE by U.S. allies or in industry that the Army can adopt?
2. What information-related practices or capabilities are being used effectively by adversaries or potential adversaries that the Army can adopt?
3. What are adversaries or potential adversaries doing in this space that U.S. Army cannot consider doing because of ethical or legal constraints but that it must be ready to counter nonetheless?

We answered these three questions through a careful examination of the doctrine and practices of selected allies, adversaries, and potential adversaries. The case studies

[6] For more, see Arturo Munoz and Erin Dick, *Information Operations: The Imperative of Doctrine Harmonization and Measures of Effectiveness*, Santa Monica, Calif.: RAND Corporation, PE-128-OSD, 2015.

focus on four allies (Israel, the North Atlantic Treaty Organization [NATO], Canada, and Germany), four state actors of concern (China, North Korea, Iran, and Russia), and four nonstate actors of concern (Hezbollah, al-Qaeda, the Islamic State of Iraq and the Levant [ISIL]/Daesh, and drug-trafficking organizations [DTOs] in Mexico).

Note that we did not formally include a U.S. Army baseline case among our case studies. Exploring U.S. concepts and capabilities at the level of detail required would be too sensitive for a general audience, as could a precise enumeration of gaps between current U.S. capabilities and those of partners or actors of concern. However, identifying the virtues of other actors' concepts and capabilities was sufficient and allowed us to make recommendations for the Army as it plans its investments in future capabilities. Other RAND and U.S. military publications provide a more focused review of U.S. IO and IRCs.[7]

A companion volume to this report presents the detailed case studies that support the conclusions and recommendations summarized here.[8] To compile these case studies, we relied on a host of secondary sources, ranging from formal published doctrine to news reports, briefings by subject-matter experts and informal chats with these experts at conferences and workshops, and white papers and articles by various research organizations.[9] Each case follows an analytic template that includes background and an overview of the case country or organization, concepts and principles relating to operations in and through the IE, how the actor organized to conduct operations in and through the IE, details on relevant operational practices, and lessons from the case for the U.S. Army. The cases as presented in the accompanying volume are not wholly comprehensive to maintain their utility as a resource for planners with a specific focus on the concepts, approaches, and capabilities that are most relevant to U.S. Army operations. In the interest of unifying the concepts discussed in support of this goal, some analyses may be influenced by our understanding of the cases and their applicability to the Army based on material and expertise that is not explicitly referenced in the case studies.

The comparative analyses presented in Chapter Three of this volume rely on fairly coarse overall assessments of the extent to which others excel in certain capability areas. These scores are holistic assessments based on the material considered for this study and represent the expert judgment of the research team. When a score indicates that a country or actor exceled in exercising a particular capability, that country or actor

---

[7] See, for example, Christopher Paul, *Information Operations Doctrine and Practice: A Reference Handbook*, Westport, Conn.: Praeger Security International, 2008; Porche, Paul, et al., 2013; and U.S. Joint Chiefs of Staff, *Information Operations*, Washington, D.C., Joint Publication 3-13, incorporating change 1, November 2014.

[8] Christopher Paul, Colin P. Clarke, Michael Schwille, Jakub P. Hlavka, Michael A. Brown, Steven S. Davenport, Isaac R. Porche III, and Joel Harding, *Lessons from Others for Future U.S. Army Operations in and Through the Information Environment: Case Studies*, Santa Monica, Calif.: RAND Corporation, RR-1925/2-A, 2018.

[9] Each of the case studies is fully documented with source citations and suggestions for further reading, all of which can be found in the references section at the end of the companion report.

demonstrated high levels of effectiveness. We have also assigned scores for capabilities that were undemonstrated but that, if employed, would likely have allowed a country or actor to achieve excellence in the case examined. We have adopted the standard English usage of *excel* and *excellence* as denoting only the quality of being outstanding or good. Excellence is explicitly relative: If everyone is good at something, then those who are excellent are truly outstanding.

Key Terms and Their Use

Throughout this report, we often use the phrase "operations in and through the IE." This language mirrors evolving thought within the U.S. Department of Defense (DoD) more broadly and is more inclusive than some alternative shorthand.[10] For example, the terms *IO* and *IRCs* capture the vast majority of current U.S. Army efforts in and through the IE, as well as their planning and coordination. However, in the cases studied, different terms of art are employed. We could call such efforts by others *IO* and match the closest IRC label to the capabilities fielded, such as *military information support operations (MISO)* for influence efforts or *offensive cyberoperations* for various cyberattack capabilities, but we intentionally avoided doing so to avoid mirror-imaging. Just because one of our cases included an influence capability, that does not mean that the actor involved organized like the U.S. Army organizes for MISO. Just because an actor conducted coordinated efforts to achieve effects in and through the IE, that does not mean that it conducted IO.

In most instances, we use either a generic term for the capability demonstrated in the case (e.g., "efforts in or through the IE," "influence capability") or we use the term employed by the case country or organization (e.g., Germany has "operational communication," and other actors in our cases conduct "media operations"). We follow this pattern even when the term parallels DoD language. For example, NATO, Canada, and Germany all have an information operations function, and all three also have psychological operations (PSYOP), whereas U.S. policy and doctrine now use MISO instead of PSYOP. When discussing these actors' employment of what U.S. personnel would understand as MISO, we use the term *PSYOP*. Some terms are names of U.S. military capability areas but are so generic as to denote that class of effort without mirroring U.S. structures, such as electronic warfare (EW) and operations security (OPSEC).

Some of the terms associated with operations in and through the IE are potentially confusing. For example, Joint Publication 3-13, *Information Operations*, defines the *IE* as "the aggregate of individuals, organizations, and systems that collect, process,

---

[10] See U.S. Department of Defense, 2016.

disseminate, or act on information."[11] This definition is elaborated to divide the IE into three dimensions: the physical dimension, which includes infrastructure, computers and other systems, wires, antennae, dishes, receivers, and other hardware; the informational dimension, which includes the content and data, including passwords and encryption; and the cognitive dimension, composed of the attitudes, beliefs, thoughts, and perceptions of those who transmit, receive, respond to, or act on information.[12] This is a solid and effective definition for the IE. However, some confusion can occur when *physical* becomes shorthand for the physical dimension of the IE, as there is the possibility of conflation with other kinds of physical spaces or features, such as physical effects in the operating environment that do not have anything to do with the IE. For example, a bomb strike is a physical event, but unless it hits a cell tower, a receiver dish, or some other piece of communication infrastructure, it has nothing to do with the physical dimension of the IE.[13]

Another potentially confusing term is *IO*. Joint Publication 3-13 defines IO as "the integrated employment, during military operations, of IRCs in concert with other lines of operation to influence, disrupt, corrupt, or usurp the decisionmaking of adversaries and potential adversaries while protecting our own."[14] This could be a synonym for operations in the IE to affect the decisionmaking of adversaries and potential adversaries, but it isn't. IO as conceived and practiced by DoD are a staff function—a coordinating, integrating, and planning function—not "operations," per se. Actual operations in and through the IE are perhaps more appropriately described as the execution of IRCs, where an IRC is "a tool, technique, or activity employed within a dimension of the information environment that can be used to create effects and operationally desirable conditions."[15]

The easy conflation of IO (a staff function) with actual operations in and through the IE (IRC execution) is akin to conflating fire support coordination and artillery. Just as one would not expect fire support coordination personnel to travel to the forward positions of artillery formations and assist in the laying of guns, one would not expect

---

[11] U.S. Joint Chiefs of Staff, 2014, p. I-1.

[12] U.S. Department of Defense, 2016.

[13] Further complicating matters, such a bomb strike might not affect the *physical* dimension of the IE, but it may very well affect the *informational* dimension (when it is reported in the battle damage assessment or photographed and posted to the Internet). As a result, it would also affect the *cognitive* dimension of the IE as people learn about, think about, and have feelings in reaction to the bomb strike.

[14] U.S. Joint Chiefs of Staff, 2014, p. I-1.

[15] U.S. Joint Chiefs of Staff, 2014, p. GL-3. The traditional core IRCs include MISO, EW, cyber operations, OPSEC, and military deception (MILDEC). Traditionally supporting or related IRCs include public affairs, civil-military operations, and special technical operations. A broader interpretation of the term (which we encourage) would extend beyond these capabilities whose effects are focused primarily on the IE to those whose primary focus may be elsewhere but can, intentionally or otherwise, generate effects in the IE. This might include any capabilities used for maneuver or fires, as well as the presence, posture, and profile of troops.

IO personnel to record a broadcast, print a leaflet, or otherwise directly participate in IRC execution. Similarly, one would never refer to "fires" as "fire support coordination."

A final term is worthy of note: *propaganda*. While this term appears in joint doctrine, it is not specifically defined. The dictionary definition of *propaganda* is "the spreading of ideas, information, or rumor for the purpose of helping or injuring an institution, a cause, or a person."[16] What is not captured in that definition is that, in American English at least, *propaganda* is highly pejorative. Invoking *propaganda* implies underhandedness, manipulation, partial truth (or complete fabrication), and an inappropriate effort to help or harm through information. Here, when we use the term *propaganda,* we include and intend this pejorative denotation.[17]

How This Report Is Organized

Chapter Two offers a brief summary of each of the others studied. Chapter Three begins the comparative analysis, first identifying areas in which the different cases studied excelled and then exploring features common to cases with significant success or excellence. Chapter Four extends the comparative analysis, focusing on distinctive and remarkable features of individual cases. Chapter Five summarizes our findings and presents recommendations for the Army. A companion report, *Lessons from Others for Future U.S. Army Operations in and Through the Information Environment: Case Studies*, presents the full case studies and detailed lessons for Army operations.[18]

---

[16] "Propaganda," *Merriam-Webster Online Dictionary*, undated.

[17] For more on the history of this term, see Christopher Paul, *Strategic Communication: Origins, Concepts, and Current Debates*, Santa Barbara, Calif.: Praeger, 2011, pp. 44–46.

[18] Paul et al., 2018.

CHAPTER TWO

Summary and Overview of the Cases

This chapter presents a brief summary of our 12 case studies, which focus on four allies (Israel, NATO, Canada, and Germany), four state actors of concern (China, North Korea, Iran, and Russia), and four nonstate actors of concern (Hezbollah, al-Qaeda, ISIL, and DTOs in Mexico). A companion report, *Lessons from Others for Future U.S. Army Operations in and Through the Information Environment: Case Studies*, presents the full case studies and detailed lessons for U.S. Army operations.[1]

Topics Covered in the Case Studies

The case studies—presented in full in the companion volume—follow roughly the same outline, with some variation in specific terms of art. Each major discussion category addresses one or more topics, determined on a case-by-case basis as relevant to or interesting about the case or according to what data were available:

- Summary
- Background and overview
- Concepts and principles for operations in and through the IE
  - Topics addressed, as applicable to the case: strategic goals/vision, how operations in the IE fit within the actor's overall strategic goals, targets and audiences, foundational principles, doctrinal principles, history and evaluation of these concepts and principles
- Organization for operations in and through the IE
  - Topics addressed, as applicable to the case: structure, funding/budget/legal authorities, doctrine, functional/organizational divisions, IRCs available, personnel, relationships and dependencies, organizations considered part of IE-related enterprise, structure for planning and integrating operations in and through the IE, coordination/integration challenges, relationship with other military or government functions

[1] Paul et al., 2018.

- Operations in and through the IE in practice
  - Topics addressed, as applicable to the case: examples of interesting operations in and through the IE, key initiatives or developments, effectiveness of efforts in the IE, efforts to counter efforts and their effectiveness, vulnerabilities in efforts
- Lessons for the Army
  - Topics addressed, as applicable to the case: operations in the IE in contrast with U.S. IO and IRCs, key takeaways from operations in the IE.

The remainder of this chapter briefly reviews each case study; Chapters Three and Four present the results of our comparative analysis of the cases.

Israel

Israel and the Israel Defense Forces (IDF) face a near-constant threat from both nation-states and violent nonstate actors in the Middle East. As such, they are compelled to test new ideas and rigorously assess both successes and failures, as well as adapt and evolve appropriately, in the IE. Israel's military has faced a sharp learning curve when operating in and through the IE. This is partly due to the distinctive emphasis placed on the main audience for its information efforts: its domestic population. This narrow focus has contributed to some critical errors and missed opportunities over time, but like any advanced force, Israel has sought to learn from its mistakes, and, since the Winograd Commission in 2008, there has been a much greater focus on coordinating communication and information efforts across the entire government to avoid information fratricide. Furthermore, Israel has worked diligently to both document and explain operations to domestic audiences, and it compellingly refutes false claims that might undermine domestic support. As such, embedded press, combat camera or equivalent, independent journalists, and other sources of transparent reporting and verification have played a vital role in insulating against falsehoods and propaganda. Finally, of all the countries in the Middle East, including its chief adversary, Iran, Israel retains the most advanced cyber capabilities, including offensive cyber capabilities and the ability to stoutly defend its own networks and critical infrastructure.[2]

NATO

NATO IO are closely integrated with the efforts of its member states, but they are rooted in distinct doctrine. NATO IO have been forced to evolve in response to

[2] Rhea Siers, "Israel's Cyber Capabilities," *Cipher Brief*, December 28, 2015.

changes in the threat environment. NATO's efforts prior to 1990 focused on deterring the Soviet Union and its influence. The emphasis shifted to the Balkans in the 1990s, the Middle East in the 2000s, and back to the European periphery in the wake of the conflicts in Ukraine and Syria. The challenges faced by the International Security Assistance Force in Afghanistan prompted the most significant efforts to improve NATO's IO-related capabilities. Since that time, NATO has ratified several key policy documents, including a strategic communication concept in August 2010, and it has been pursuing better cooperation between its international military staff and its members' national armed forces. In 2016, a new Assistant Secretary-General for Intelligence post was proposed in a clear attempt to address the threat that emerging powers pose in the information space. Despite its limited resources and dependence on individual member contributions and unanimity-based decisionmaking, the alliance's integrative function has never been in higher demand to address threats emanating from state and nonstate actors alike.

Canada

Canada positions itself to its home audience as a global partner—specifically, featuring itself in a stabilization operations role. The focus is mainly on interoperability with partners rather than constructing a narrative of Canadian force dominance. The Canadian Armed Forces (CAF) have a series of IRCs, or "enablers," in their arsenal, though synchronizing IO remains a work in progress.

The functional model of the Canadian military maintains IRC personnel as part of the reserve component, with the CAF unable to use the capabilities until relevant personnel are funneled into a state of high readiness and called to support a specific mission. There is no standing command dedicated to IO, so enablers tend to be integrated on a situational basis. In many cases, they are Canadian Reserve Force personnel who are trained in the IRC but do not have a specific related military occupational specialty. However, there has been progress in developing IRCs across the CAF; for example, the Influence Activities Task Force is standardizing its training, planning, and doctrine.

There has also been progress in synchronizing PSYOP with civil-military cooperation in colocated influence activity companies, but the CAF have yet to achieve full strategic coordination with counter-command activities and information protection activities, leaving the importance and fit of IO for Canadian national missions undefined. Beyond a commitment to readiness training, it is unknown whether the current operating model adequately aligns with the execution of strategic IO goals. The CAF have not deployed their influence activity structure at the company level in more than five years. Cyberspace is beginning to receive increased attention in Canada and could provide an opportunity for Department of National Defence officials to reassess CAF organizational priorities. There is some evidence that the integration of IO enablers may be on the horizon, though without a high-level champion to advocate for synchronization, it is questionable whether CAF IRC enablers will be adept enough to function effectively in the quickly evolving IE.

Germany

Since Germany's reunification in 1990, its military (Bundeswehr) has developed a unique and well-resourced capacity to engage local populations in areas of deployment and to communicate information to its troops. These two activities have been the key focus of the Center for Operational Communication of the Bundeswehr, founded in 2015, and the several battalions designated for psychological warfare before that. Given its complex history, however, Germany has a relatively narrow definition of military information: It has completely refrained from using misleading or selective information and employs strict controls to ensure that information disseminated both to its troops and to local populations is as accurate and complete as possible. The Bundeswehr's ability to counter foreign propaganda is underdeveloped, particularly in light of how the security environment has evolved in Europe since Russia's invasion of Georgia in 2008. However, the German military is constantly evaluating the effectiveness of IO and its Center for Operational Communication. Electronic warfare and cyberwarfare capabilities are not part of German operational communication but are housed under joint staff elements. Germany has one of the three best-resourced IO (or PSYOP) capabilities among NATO member states, alongside the United Kingdom and the United States, and it regularly engages in training allied forces. In their foreign deployments, German IO personnel are typically embedded in combat and other units and rely heavily on local stakeholders to disseminate information and win public support for German strategic objectives on the ground. Similar to other European countries, Germany has effectively blurred IO and public affairs (PA), although some important distinctions remain.

China

The People's Republic of China is actively and aggressively pursuing goals to persuade the global population of its resurgence while limiting access to information at home in an effort to maintain a one-party system.[3] To do this, China has aggressively entered the IE, systematically developing capabilities and then exercising them in real-world operations. Indeed, over the past 25 years, it has closely followed the

---

[3] Reporters Without Borders, "China," web page, undated.

actions of the United States in the IE; it has taken these concepts, blended them with Soviet-era doctrine, and added its own traditional tactics and concepts to create a uniquely Chinese approach to information warfare (*xinxi zhanzheng*). These efforts are aimed at global audiences, internal audiences, and the West, particularly the U.S. public and policymakers. Many of the operations Beijing is pursuing are designed to sway public opinion, acquire U.S. technologies, and counter U.S. action throughout the IE. It wants to do this without physically engaging the United States in open conflict, and it is developing its IRCs, doctrine and policy, force structure, and legal frameworks—as well as adding resources—to accomplish this goal. Indeed, China and, by extension, the People's Liberation Army (PLA), is already actively engaged in multiple "warfares"—psychological, public opinion, and legal—even though there is no active conflict.[4] The PLA has gone through a massive reorganization in which it has consolidated all IO and IRC forces into the Strategic Support Force, which it has designated a fifth service branch, indicating its high level of importance. China also engages in civil-military operations and civil affairs–like projects, especially in parts of Latin America and Africa, where Chinese investments in infrastructure projects are intended to curry favor with government leaders and domestic populations.[5] While the United States recognizes that certain PLA actions are designed to bolster regional territorial claims and diminish U.S. influence in the region, it is not fully prepared for the suite of capabilities that the PLA is currently employing. This case study examines China's actions, capabilities, and developments with the aim of raising Army leadership awareness.

North Korea

More than 60 years of isolation and inflammatory posturing have left the North Korean regime with limited options to influence regional and global actors. It relies on nuclear and ballistic missiles and a huge standing army, along other asymmetric capabilities—such as electronic warfare, cyber warfare, and information warfare—to deter aggression from outside entities. Coupled with its leadership's bellicose rhetoric, North Korea's deterrence strategy is designed to intimidate other countries and maintain regime control.

---

[4] Dean Cheng, "Winning Without Fighting: The Chinese Psychological Warfare Challenge," Washington, D.C.: Heritage Foundation, Backgrounder No. 2821, July 11, 2013.

[5] For more on Chinese involvement in Africa, see Lloyd Thrall, *China's Expanding African Relations: Implications for U.S. National Security*, Santa Monica, Calif.: RAND Corporation, RR-905-A, 2015, and Larry Hanauer and Lyle J. Morris, *Chinese Engagement in Africa: Drivers, Reactions, and Implications for U.S. Policy*, Santa Monica, Calif.: RAND Corporation, RR-521-OSD, 2014. For background on Chinese investments in Latin America, see David Dollar, *China's Investment in Latin America*, Washington, D.C.: Brookings Institution, Geoeconomics and Global Issues Paper 4, January 2017.

The regime focuses its efforts on influencing two audiences: (1) external actors, including its regional neighbors, the West, and the international community writ large, and (2) groups inside North Korea, including regime elites, the Korean People's Army, and the North Korean public. The regime is specifically worried about dissent from its domestic audience and therefore devotes significant effort to controlling access to information and creating a god-like "cult of personality" around supreme leader Kim Jong-un.

Iran

Iranian influence operations are directed at multiple audiences. Domestically, the goal is to promote the view that the country is unfairly targeted by the rest of the world, particularly the United States and Israel. Across the broader region, Iran's efforts in the IE are largely directed toward its ongoing proxy conflict with Saudi Arabia, in which Tehran supports Shia populations in Iraq, Yemen, Lebanon, Syria, Bahrain, and elsewhere. The Iranian Revolutionary Guard Corps Quds Force is an elite military unit used by the regime to train proxy forces abroad as a means of extending Iranian foreign and security policy across the Middle East.

On a more fundamental level, Iran is beginning to devote a significant portion of its resources, attention, and energy to cyberspace. It has employed cyber capabilities less to exercise real aggression than to signal its capability and to act as a deterrent, which involves both having punitive capabilities and demonstrating a resolve and willingness to use them. Iran has used technical IRCs in exactly this way. Iranian hackers have progressed far beyond website defacing or distributed denial-of-service attacks; they are now capable of developing sophisticated software to probe U.S. systems for vulnerabilities, inject malware, and gain control. In its operations in and through the IE, Iran has no compunction about disseminating falsehoods or manipulating information and relies extensively on informational combat power in operations short of war. Iran engages in civil-military operations and civil affairs–like activities through its patronage in the Middle East, but especially in Syria since the beginning of the civil war there in 2011.[6]

Russia

Engaging opponents in the IE is not part of a forgotten Cold War past in Russia. Rather, it is a key component of its current foreign *and* domestic policy. The past few

---

[6] Ahmad Majidyar, "Celebrations of Iranian Revolution Across Syria Shows [sic] Iran's Soft Power Hegemony," Washington, D.C.: Middle East Institute, February 13, 2017.

years have illustrated the breadth and depth of the Russian information apparatus, which the state has nimbly adapted to the digital age with significant investment in efforts to influence public opinion via the Internet and in foreign countries. Remarking on the capabilities that Russia has built and its willingness to deploy them domestically and internationally, General Philip Breedlove bluntly referred to the country's current efforts in the IE at the 2014 NATO Summit in Wales as "the most amazing information warfare blitzkrieg we have ever seen in the history of information warfare."[7]

Despite its economic woes, Russia has reinforced its military by developing rapid-reaction, highly deployable units and, with that, well-trained information warfare staff and a surrounding infrastructure of state-controlled media, tight controls on information shared with the public, and a brutal strategy of intimidating and silencing its critics.[8] These developments pose significant challenges for the United States and its allies, which have focused on avoiding military confrontation on NATO's Eastern flank and limiting the influence of Russia and its proxies among their own populations.

Hezbollah

Hezbollah is a Shia terrorist organization based in Lebanon and supported by Iran that has been active since the early 1980s. Over the past several decades, Hezbollah has professionalized its military force and developed a comprehensive appreciation of the value of effects in and through the IE. Hezbollah's IRCs have narrowed the divide between what was once a parochial militia and the region's most powerful military, the Israel Defense Forces. Among violent nonstate actors, Hezbollah is certainly the first group to truly appreciate the importance of an "information-first" operational mindset, which has undoubtedly contributed to its longevity and tactical success. Hezbollah is important for the U.S. Army to consider because it is an example of what violent nonstate actors can achieve in the IE by investing substantial resources in media, propaganda, and the integration of IRCs into a broader defense architecture.

Al-Qaeda

As the perpetrator of the deadliest terrorist attack in history on U.S. soil, al-Qaeda has demonstrated an intuitive understanding of the information value of kinetic operations. "Spectacular" attacks like those launched against New York City and Washington, D.C., on September 11, 2001, have echoes in the IE that equal—or even

---

7 Peter Pomerantsev, "Russia and the Menace of Unreality: How Vladimir Putin Is Revolutionizing Information Warfare," *The Atlantic*, September 9, 2014.

8 Andrew E. Kramer, "More of Kremlin's Opponents Are Ending Up Dead," *New York Times*, August 20, 2016.

surpass—the actual physical effects. Al-Qaeda's media production is sophisticated, both aesthetically and historically. However, its messages are often wide-ranging and unfocused. Since 9/11, core al-Qaeda's propaganda has abated significantly, and it is now more appropriate to think of the group's operations in and through the IE in terms of its franchises and affiliates (e.g., al-Qaeda in the Arabian Peninsula, al-Qaeda in the Islamic Maghreb, al-Shabaab).

Although its propaganda has slowed down, it lives forever on the Internet. ISIL has clearly learned from, improved upon, and surpassed al-Qaeda's tactics, which included footage showing attacks on U.S. troops, al-Qaeda militants assembling improvised explosive devices, and suicide bombers' martyrdom tapes, complete with anti-American and anti-Israeli vitriol. Deterring, disrupting, or destroying the physical organization does not put an end to the influence that a group can have, as evidenced by the popularity of Anwar al-Awlaki, whose YouTube sermons have inspired terrorist attacks long after his death. Media distributed by violent nonstate actors, such as al-Qaeda, can reinforce the group's strategy while also having a more tactical effect (e.g., conveying instructions for constructing homemade bombs in *Inspire*, the manual that Dzhokhar and Tamerlan Tsarnaev used to build the bombs for the Boston Marathon attack).

ISIL/Daesh

Since ISIL stormed through parts of Iraq and Syria in the summer of 2014, the group has cultivated a lure of invincibility among terrorism researchers and policymakers.[9] To be sure, much of what has been said about ISIL amounts to hyperbole: The militants are far from omnipotent, as witnessed by the coalition's recapture of territory in Fallujah, Ramadi, Manbij, and other critical ISIL strongholds. However, one area in which ISIL has indeed lived up to its reputation is in its ability to operate in and through the IE. ISIL has been successful in the IE for several reasons. First, information personnel are accorded high levels of prestige or are otherwise well rewarded. Second, the caliphate narrative is incredibly effective, for both unifying the group's operations and messages and providing a compelling frame to supporters and potential supporters. Third, the group's major themes are cleanly grouped and tightly focused, making message discipline easy. The themes are also directly related to several important and diversified subnarratives that specifically target different audiences.

ISIL has taken advantage of social media to disseminate its message and ideology far beyond what al-Qaeda was ever able to achieve.[10] Despite the attention afforded

[9] ISIL is also known pejoratively as Daesh, the Arabic acronym for the group's former name, the Islamic State in Iraq and Syria.

[10] Bennett Seftel, "What Drives ISIS," *Cipher Brief*, May 5, 2016.

to its execution videos, ISIL actually produces much more material, and on a broader range of topics, than what gets reported in the mainstream media.[11] ISIL propaganda is centered on three major themes. First, it has restored the caliphate, which makes it the only authentic Islamic state in the world and thus worthy of political legitimacy.[12] Second, any existing Islamic entity (state or nonstate) that does not recognize the group's authority qualifies as an apostate and must be vanquished. Finally, ISIL is more capable than al-Qaeda ever was and continues to grow as an organization and as an ideology.[13]

Mexican Drug-Trafficking Organizations

Mexican DTOs have made extensive use of actions with effects in and through the IE, especially since the surge of violence that struck Mexico in the mid-2000s. DTOs operate in constant conflict with Mexican law enforcement, rival gangs, and the threat of vigilante groups of citizens in occupied territories. They engage in profit-seeking criminal ventures in addition to drug trafficking, including extortion. In this environment, DTOs thrive on reputation and influence, and they go to elaborate lengths to build their own reputations and to slander the government, civil society, and rival organizations.

Violence is frequently both the medium and the message. Public statements, such as narcobanners (cloth banners bearing warnings, threats, or messages of accountability) and corpse messages (where the way in which a body is displayed or the wounds or other indignities to which it has been subjected send a specific message), are a critical means for a group to broadcast its intentions and demonstrate that it has followed through on them. Although Mexican DTOs lack the complex infrastructure or organization for producing and disseminating media, their acts of violence and intimidation are nonetheless carried out in ways that allow traditional and social media reporting to enhance, extend, and echo their information effects.

Mexican DTOs strive to call attention to events that would normally be considered morally repugnant. However, in the broader context of narratives and organizational mission statements that DTOs craft and disseminate, the public display of atrocity is foundational to an organization's reputation and, in turn, its ability to turn profits and protect against existential threats. These narratives vary by group. Some

[11] Aaron Y. Zelin, "Picture or It Didn't Happen: A Snapshot of the Islamic State's Official Media Output," *Perspectives on Terrorism*, Vol. 9, No. 4, 2015.

[12] Colin P. Clarke and Chad C. Serena, "To Defeat ISIL's Brand, Its Territory Must Be Reclaimed," *National Interest*, July 8, 2016.

[13] Daveed Gartenstein-Ross, Nathaniel Barr, and Bridget Moreng, "How the Islamic State's Propaganda Feeds into Its Global Expansion Efforts," *War on the Rocks*, April 28, 2016.

have crafted elaborate quasi-religious belief systems; others merely claim to be the lesser of competing evils.

CHAPTER THREE

Comparative Analysis: Common Themes in the Cases

Accepting that the IE poses a wide range of challenges and that the United States can learn from how others have addressed these challenges leads us to look for common threads and effective practices across the cases. This chapter provides a comparative overview of all the cases studied. We first describe which cases excel in which capability areas to offer a refined summary of who is good at what. We then consider common themes, concepts, and principles across the different cases, describing which of those characteristics appear to consistently contribute to effectiveness and excellence. Collectively, these characteristics can be contrasted with current U.S. Army practices and capabilities to suggest new concepts for exploration and capability areas for investment.

Capability Areas in Which Others Excel

To learn from others, we need to first identify areas in which those others are successful and then identify practices, principles, and approaches that have led to those successes. This section addresses that first topic, summarizing the capabilities areas that formed the structure for this study and the relative excellence of the different case actors across those capabilities.

Table 3.1 summarizes the capability areas in which each actor was more or less successful or capable. The rows list the cases, while the columns list the capability areas. Most of the capability areas are generic versions of (and map directly to) specific U.S. IRCs. We chose not to use IRC labels because the capability areas as conceived and employed by others may be slightly different from those of the United States, and we risk unproductive mirror-imaging if we impose U.S. concepts and terminology on other actors' activities.[1] Where a case has unambiguously excelled in a capability area, it received a check (√). Where a case had strong but not truly exemplary capability in an area, it received a ½. Blank cells denote limited capability or accomplishment in the capability area or indicate that the capability is unremarkable when compared with other cases. As described in Chapter One, these assessments are based on our holistic

[1] For more on this, see the discussion of terms in Chapter One.

Table 3.1
Capability Areas in Which Others Have Excelled

Case	Excels in press relations/media operations/public relations/government broadcasting	Excels in influence	Excels in leveraging narrative	Excels in deception/stratagem/manipulation	Excels in electromagnetic spectrum operations/electronic warfare/jamming	Excels in cyberoperations	Excels in OPSEC/secrecy/denial/information security	Excels in censorship/information control	Excels in the use of maneuver and fires as IRCs	Excels in outsourcing/use of proxies or militias/franchising for IRCs	Excels in leveraging social media/new media/now media
Allies											
Israel					½	√					
NATO							½				
Canada							½				
Germany	√				½		√				√
State actors of concern											
China	√	√	½	√	√	√	√	√	√	√	√
North Korea			√			√	√	√			
Iran						√	½	½		√	
Russia	√	√	√	√	√	√	½	√	√	√	√
Nonstate actors											
Hezbollah	√	√	√				½		√		√
Al-Qaeda							½		√		√
ISIL/Daesh	√	√	√				½	½	√	√	√
Mexican DTOs		√	√					½	√		

KEY:
Blank = not doing/not significantly present/not excellent compared with other cases
√ = significantly present in or characteristic of case
½ = partially present in or somewhat characteristic of case

understanding of all the cases taken together. The companion volume provides additional detail supporting these assessments.[2]

A cursory examination of Table 3.1 reveals that several cases demonstrated noteworthy capability in every aspect of operations in the IE, but only China and Russia have a check or ½ in every column. Among the nonstate actors, Hezbollah and ISIL are noteworthy, having excellent capabilities in more than half of the listed capability areas. Among U.S. allies, Germany is a standout, though with fewer capability areas of excellence than almost all the actors of concern. Different cases demonstrate different areas of excellence, but all cases had at least some areas of excellence. Furthermore, even cases with few areas of excellence can contribute valuable lessons (as we highlight in the next chapter and in the accompanying case studies). For the comparative analyses, we extracted general lessons and sought to identify practices and concepts present in the cases that were consistently (or at least predominantly) excellent in the identified capability areas. It may seem anomalous that U.S. allies tend to demonstrate fewer areas of excellence than the various actors of concern. This is consistent with our overall assessment, but is not as simple or straightforward as it might seem. While part of the reason for this unfavorable comparison stems from the relative emphasis of efforts in and through the IE (both in spending and in priority during operations), it stems in part from ethical and legal constraints that allies place on their efforts, which the various actors of concern largely do not. Simply put, it is easier to excel when you are willing to cheat, and pointing out that U.S. allies lag in certain areas should not be construed as a recommendation that they abandon their ethics and principles. Rather, both the United States and its allies need to recognize where and how actors of concern gain advantage and seek to minimize the advantage they gain from behavior outside of Western norms and values.

Several cases demonstrate excellence in press relations, media operations, public relations, or government broadcasting. Specifically, the four highlighted actors of concern that demonstrate the most comprehensive excellence (Russian, China, Hezbollah, and ISIL), as well as U.S. ally Germany, excel in this capability area. This capability area is most closely akin to the traditional PA IRC, though it also touches on and encompasses government broadcasting, a capability not traditionally considered part of the U.S. defense portfolio.

Excellence in influence corresponds most closely to military information support operations (MISO, formerly psychological operations). However, while MISO is confined to targeting selected foreign audiences, many of the actors of concern employ their influence capabilities to influence their domestic audiences, too.

Related to excellence in influence is excelling at leveraging narrative. This capability area does not align clearly and unambiguously with an existing DoD IRC, something that, itself, suggests a gap. By *leveraging narrative*, we mean to denote excellence

---

[2] Paul et al., 2018.

in using storytelling to convey information (and generate influence), or motivating a call for action, as well as taking advantage of (and sometimes modifying) existing stories and narratives to enable and support the accomplishment of other objectives in and through the IE. In other research, we have identified three particularly relevant characteristics of narrative: (1) it is used to make sense of the world and one's place in it; (2) compelling narratives employ consistency, familiarity, and proof; and (3) narratives already exist, and although they can be shaped over time, they cannot always be changed or replaced.[3]

Excellence in deception, stratagem, or manipulation as a capability area is most closely aligned with MILDEC. This capability area includes both Russian *maskirovka* (camouflage and obfuscation in all things) and the more insidious Russian concept of "reflexive control," in which forces convey selected information to manipulate the decisions of their foes, without those foes ever becoming aware that they were manipulated.[4] The Chinese have emphasized the value of stratagem for centuries, back to and including Sun Tzu's famous aphorism that "all warfare is based on deception."

Excellence in electromagnetic spectrum operations, EW, and jamming is most closely aligned with electromagnetic spectrum operations and EW. Excellence in this relatively straightforward technical capability area simply requires significant investment in technology and related systems and a willingness to employ them aggressively. Because of resource and technology constraints, none of the nonstate actors excelled in this area, while Russia and China exhibited significant capabilities.

Excellence in cyber operations as a capability area aligns with U.S. cyber-related capabilities. Note that true excellence in this capability area requires a level of investment in technology that makes it the exclusive provenance of state actors, though nonstate actors can certainly have capabilities in this area (both organic and provided by state actors wishing to act through a proxy).

Excellence in operations security, secrecy, denial, or information security aligns most closely with OPSEC and information assurance. Most of the actors we studied were accomplished in this area. For nonstate actors, OPSEC and information security can be imperative to survival, as these actors have persistent adversaries actively hunting them. Yet, all the nonstate actors in our cases exhibited a willingness to sacrifice low-level officers by diverting resources to the protection of key operatives. State actors recognize the imperative of keeping plans and intentions secure during times

---

[3] Christopher Paul, Kristen Sproat Colley, and Laura Steckman, "Fighting Against, with, and Through Narrative," *Marine Corps Gazette*, forthcoming.

[4] Timothy Thomas, "Russia's Military Strategy and Ukraine: Indirect, Asymmetric—and Putin-Led," *Journal of Slavic Military Studies*, Vol. 28, 2015, p. 456; *Reflexive control* was "developed and used during the Soviet era" and "has generally been understood as a means of conveying specially prepared information to a partner or an opponent to incline him to voluntarily make the predetermined decision desired by the initiator of the action." Use of the concept in Russia dates back to at least the 1960s.

of conflict, and nonstate actors with a history of information control have a leg up in this area.

Excellence in censorship or information control as a capability area is not akin to any U.S. IRC, nor is it a capability area in which the United States should seek to invest, as it is inconsistent with U.S. democratic values. Although changes in the IE have made information control more difficult than it was in the past, several state actors maintained excellence in this area through significant constraints on the Internet within their borders, control of technology, state ownership (or sympathetic ownership) of various media outlets, and stridently enforced censorship and media-control statutes and regulations.

Excellence in the use of maneuver and fires as IRCs indicates a recognition that "one cannot *not* communicate" and that every action, utterance, and expression of every soldier, sailor, airman, and Marine in an area of operations sends a message.[5] This includes what the Australians call presence, posture, and profile—the messages implicit in where troops are and how they comport themselves.[6] This is just the logical extension of the same sort of reasoning that supports "gunboat diplomacy" and all the traditional approaches to deterrence, which use physical capabilities to realize effects in and through the IE.

Excellence in this capability area involves the integrated employment of information and physical power to achieve informational or cognitive objectives. This is certainly the domain of nonstate actors, all of which were asymmetrically disadvantaged when matching physical capabilities against physical capabilities in our cases, so they instead sought to create an asymmetric advantage for themselves in and through the IE. Both Russia and China also excel in this area, using maneuver and fires as an integrated part of their deceptions and obfuscations, along with the physical posture of their forces and platforms as part of their intimidation and bullying, among other things.

Excellence in outsourcing, the use of proxies and militias, or franchising for IRCs has no direct analog in U.S. operations, but is related to other capabilities and categories of activity. Both unconventional and irregular warfare allow the use of proxy forces, and various U.S. security cooperation efforts and efforts to build partner capacity can involve working "by, with, and through" partner forces, which can be akin to outsourcing or using proxies. Security cooperation that involves outsourcing for IRCs would belong in this specific capability area. The use of proxies has several potential advantages. For example, it can increase the capacity of a capability (amplification

---

[5] The first of five axioms is listed in Paul Watzlawick, Janet Beavin-Bavelas, and Don D. Jackson, *Pragmatics of Human Communication: A Study of Interactional Patterns, Pathologies and Paradoxes*, New York: W. W. Norton and Company, 1967, chapter 2. For more, see Paul, 2011.

[6] James Nicholas, "Australia: Current Developments in Australian Army Information Operations," *IO Sphere*, Special Edition 2008.

through extra personnel), it can enable plausible deniability (oh, it wasn't *us*, it was *them*), and it can decrease costs (having someone else do something might be cheaper than doing it yourself, and teaching someone to fish will certainly be cheaper in the long term than providing them with fish).

Excellence in leveraging social media/new media/now media is related to U.S. capabilities for PA, MISO, and intelligence without falling strictly within any one of those capabilities. This capability area involves the exploitation of social media as a source of data for intelligence, monitoring, or evaluation and also as a channel for messaging and influence. Excelling in this area requires three things: (1) the willingness to engage using these media, (2) permission or authority to engage using these media, and (3) investment in relevant human capital and supporting tools. The costs of such investment are relatively modest, which is why nonstate actors are able to achieve excellence in this area. Germany has also done so by making the necessary investment in human and technical capital, though with authorities that constrain engagement to certain audiences and for certain purposes.

Common Concepts and Principles Across the Cases

As discussed in the previous chapter, there is a range of different IE practices and approaches in the cases we studied. Where are there common themes in terms of concepts and principles? Which are associated with effectiveness in the IE? Table 3.2 lists concepts and principles employed across multiple cases. The rows list the cases studied, and the columns are the repeated and relevant concepts and principles. Where a concept or principle was significantly present or characteristic of a case, it received a check (√); ½ indicates that it was partially present or somewhat characteristic. Blank cells denote the absence of the concept or principle or indicate that it was not significantly present when compared with the other cases. Table 3.2 is particularly illuminating when considered in light of the observations in Table 3.1.

Reviewing Table 3.2 beginning with the first column and recalling the observations in Table 3.1, we see that one of the features present in each of the most successful cases is that information power capabilities were generously resourced. Cases featuring a heavy investment in IRCs (or equivalents), either in absolute terms or relative to investment in other capabilities (such as troops, artillery, armor, aviation, or nuclear forces), feature actors that were also most effective in the IE. This includes all the state actors of concern and three of the four nonstate actors. All were generally and broadly effective in the IE and adhered to this principle, including China, Russia, Hezbollah, and ISIL. Only one of the studied U.S. allies even partially embodied this principle: Germany is considered an exemplar of IRC strength and excellence within NATO, but its level of investment falls short of that made by all four state actors of concern.

Table 3.2
Common Concepts and Principles Governing Others' Operations in the IE

Case	Information power capabilities generously resourced	Information effects given priority/prominence in operations	Information power–related personnel accorded prestige, otherwise well rewarded	IRCs collected together in single organization or command	Integration of informational and physical power	Extensive use of informational combat power in operations below the threshold of conflict	Employed concept of getting target to unwittingly choose preferred course of action	Significant focus of information efforts on own domestic/internal audience	Careful recording/documenting of own operations	Complete commitment to "white" information—no influence or manipulation	No compunction about falsehood or manipulation	High production values for video, digital, and other products
Allies												
Israel								√	√			
NATO								½		√		½
Canada				½								
Germany	½	½		½	½			½	√	√		√
State actors of concern												
China	√	√		√	√	√	√	√			√	√
North Korea	√				√	√		√			√	
Iran	½	½			½	√		√			√	
Russia	√	√			√	√	√	√			√	√
Nonstate actors												
Hezbollah	√	√	√		√	√		√	√		√	√
Al-Qaeda	½	√			√	√					√	½
ISIL/Daesh	√	½	√	½	√	√		√			√	√
Mexican DTOs		√			√			√			√	

KEY:
Blank = not doing/not significantly present/not excellent compared with other cases
√ = significantly present in or characteristic of case
½ = partially present in or somewhat characteristic of case

Another principle common among successful cases is the priority/prominence given to information effects in operations. Whether it is Hezbollah's axiom that "if you haven't captured it on film, you haven't fought" or Russian emphasis on *maskirovka* and reflexive control, examples abound of case actors achieving their objectives solely through the use of informational power—or through a blended use of informational and physical power in which physical power often contributed to informational effects and with the informational effects being the primary effort. This is akin (in DoD-centric terminology) to not only allowing IRCs to be the main effort or the *supported* effort (rather than the *supporting* effort) but for that to be the case much of the time. One particularly striking example of information effects being given priority in operations is the case of the Mexican DTOs, in which the violence and the message were always intimately intertwined. Not only is an act of violence the elimination of a rival or the punishment of a defector, it is also a deterrent against future defection and a statement of intimidation and assertion of control.

The two nonstate actors that have enjoyed the most success in their operations in the IE (Hezbollah and ISIL) employ a practice that contributes to their success: Their information power–related personnel are accorded a high level of prestige or are otherwise well rewarded. In ISIL-held areas of Iraq and Syria, for example, those with skills appropriate for the media unit were paid upwards of 30 times the rate of a typical foot soldier, and they were accorded the rank of "emir."[7] High pay and a high degree of respect for information power personnel helps reinforce the priority given to such capabilities in operations, and it also helps with recruitment and retention of skilled personnel in these positions.

Another practice that was slightly less common in our cases but beneficial to effectiveness when it did appear is the centralization of IRCs into a single organization or command. China, for example, recently launched the equivalent of a separate military service with a primary emphasis on information-related effects and capabilities, and Germany has drawn together all of its non-technical IRCs into one formation under the rubric of operational communication. This approach can serve as a tangible demonstration of a commitment to resourcing and to the importance of information power, and it can have several benefits, including

- increasing cross-functional understanding across aggregated capabilities
- smoothly integrating technical (e.g., spectrum, EW, cyber) and content-focused (e.g., influence, public relations) capabilities
- increasing opportunities for synergy and ease of coordination across aggregated capabilities
- centralizing personnel for relevant and cross-functional training

---

[7] Greg Miller and Souad Mekhennet, "Inside the Surreal World of the Islamic State's Propaganda Machine," *Washington Post*, November 20, 2015.

- increasing the highest level of grade structure associated with the combined formation above what would be available in separate capability areas[8]
- increasing the likelihood of IRCs being "the main effort"
- increasing access to resources
- enabling the possibility of specialized recruitment efforts.

The principle of integration of information and physical power is obviously essential for excellence in the use of maneuver and fires as IRCs, but adherence to this principle has other benefits as well. In our cases, excellence in several other capability areas was bolstered by the support of both physical and informational power—specifically, influence and deception/stratagem, two particularly important capability areas. All the actors of concern, both state and nonstate, adhered to this principle to at least some extent, while U.S. allies embodied it to a much lesser extent. This points to an opportunity for the United States and its allies; unlike some of the other principles and practices embraced by actors of concern, there are no ethical barriers to the integration of physical and informational power. The U.S. Army and U.S. allies can all increase their adherence to this principle and incorporate it into operating concepts to enable effects in and through the IE without legal or ethical concerns.

Gray-zone aggression by the state actors of concern and radicalization and extremist violence by the nonstate actors are two of the major threats in our case studies.[9] Both of these activities fall below the threshold of traditional conflict and involve numerous opportunities to create effects in and through the IE. It should come as no surprise that all the actors of concern in our cases (save for the drug traffickers in Mexico) have made extensive use of informational combat power in operations below the threshold of conflict.

The two state actors of concern that were most successful in and through the IE in our case studies convinced a target to unwittingly choose the actor's preferred course of action. For Russia, this is the notion of reflexive control, briefly described earlier. For China, it is the observation attributed to Mao Tse-Tung: "To get someone to do something for himself that he thinks is in his own interests, but which is actually in your interests, is the essence of strategy."[10] In both low-intensity and high-intensity competitions or conflicts, employing this principle has the potential to substantially affect the outcome. For example, if Russia's use of obfuscation and disinformation during the invasion of Ukraine was intended to paralyze the United States and NATO into

---

[8] In a single organization, all IRCs might be headed by a senior general or flag officer; if left in capability stovepipes, each might be separately headed by a less senior general or flag officer or an O-6.

[9] For more on gray-zone aggression, see Michael J. Mazarr, *Mastering the Gray Zone: Understanding a Changing Era of Conflict*, Carlisle Barracks, Pa.: U.S. Army War College, Strategic Studies Institute, 2015.

[10] Timothy L. Thomas, "Asia-Pacific: China's Concept of Military Strategy," *Parameters*, Vol. 44, No. 4, Winter 2014–2015.

indecisive inaction, it achieved significant success. While Russia was not challenged militarily by the United States or its allies, the economic sanctions it failed to avoid had significant consequences for the Russian economy.

The majority of our cases involved information efforts with a significant focus on domestic or internal audiences. Although the coding in Table 3.2 does not make a distinction, there are two very different instantiations of this practice. The first is the version adopted by U.S. allies, in which the domestic audience is kept well informed of military missions and operations. The goal is to maintain support for those operations, but such support is secured only through the provision of true information, not through any effort to manipulate or deceive the domestic audience (which, in most of these countries, would be illegal). Then there is the second version of domestic focus, adopted exclusively by actors of concern. It involves efforts to pacify that audience or otherwise maintain support for military operations or adventures *through any means necessary*. This can involve a host of deceptions and manipulations, including stoking the fires of nationalism, falsely blaming others for provocations, or mobilizing (or changing) religious, ethnic, sectarian, or cultural narratives to promote support (even at the expense of exacerbating existing divisions and tensions).

A practice that also serves two different purposes and has two different versions is the careful recording or documenting of one's own operations. Both Israel and Germany extensively document their own operations to inform their domestic constituencies and to refute the propaganda of others. In addition, both of these countries use historical lessons learned to rethink future operations, and Germany has been particularly proactive in restructuring its IO capability to better provide an integrative and coordinating function within the Bundeswehr. Hezbollah carefully documents (usually though video imagery) its operations for exactly the opposite purpose: to create propaganda and to make others look bad.

Many U.S. allies adhere to the principle of complete commitment to "white" information—that reported without intent to influence or manipulate. White information is completely truthful both in its content and attribution. Gray or black information is some combination of incomplete or partially false content or incomplete or partially false source attribution; that is, it relies on one omitting the identity of the source of the information or misleading about the identity of the source. The strict adherence to white information is very much a two-edged sword. On the one hand, it does improve the credibility of a force and government, and it completely eliminates any internal organizational or statutory limitations that might separate which content or capabilities can be directed at a domestic audience and which are suited only for selected foreign audiences. Commitment to exclusively white information maximizes the prospect for positive relations with domestic audiences in countries where the government is unwilling to propagandize its own constituents. On the other hand, such a commitment completely precludes the possibility of efforts to aggressively influence adversary forces, decisionmakers, or supporting populations during times of conflict or intense competition. While the complete adherence to white information is ethically appealing and does have some benefits, a review of Figure 3.1 makes clear that it is not associated with a high level of effectiveness across numerous IRCs.

In contrast, all of the actors of concern in our cases had no compunction when it came to employing falsehood or manipulation. While one might be tempted to assume that this disadvantaged these actors in some way, the disadvantages of falsehood are far less severe than one might expect. Research in psychology shows that people are poor both at discriminating the credibility of sources and information and at recalling those discriminations. Thus, propaganda campaigns that involve falsehoods are not particularly disadvantaged relative to truth-based influence efforts, and they may have some advantages.[11]

Finally, several cases, including all of the most generally successful, involved video, imagery, and other information products with high production values. High-quality video and imagery, skilled voice acting with accents perfectly matched to target audiences, and slick sets and backgrounds all contribute positively in the capability areas in which they are used—and nonstate actors often leverage the low cost of producing such material relative to other military operations.

Key Takeaways

This discussion highlighted several practices and principles worth emulating and several more practices and principles that are sinister and cannot be emulated but must be prepared for. Practices and principles worth emulating include the following:

- generous resourcing of information power and IRCs
- giving prominence to information effects in operational planning and execution (including expanding authorities for their use)
- increasing the prestige and regard of IRC personnel (through, for example, recruiting incentives and other incentive pay)
- integrating physical and information power
- extensively using information power in operations below the threshold of conflict
- employing the concept of getting the target to unwittingly choose one's preferred course of action
- carefully recording and documenting one's own operations, and enabling embedded press, independent journalists, and combat camera (or an equivalent) as witnesses in the battlespace
- high production values.

[11] Christopher Paul and Miriam Matthews, *The Russian "Firehose of Falsehood" Propaganda Model: Why It Might Work and Options to Counter It*, Santa Monica, Calif.: RAND Corporation, PE-198-OSD, 2016.

Actors of concern operate with a number of advantages in the IE. One is that they are unconstrained by ethics, law, or policy; as an old IO hand once asked the lead author, "How do I compete with a head on a stick?" The point was not that he wanted to be able to put heads on sticks. It was that the adversary could and would do almost anything (and put almost anything on the Internet) to grab attention and communicate its message. Such actors will continue to employ practices and principles that are unacceptable under U.S. values yet effective in the IE, including the following:

- a focus on maintaining the support of domestic audience by any means necessary
- a lack of compunction in spreading falsehoods or engaging in manipulation
- censorship and information control.

CHAPTER FOUR

Comparative Analysis: Distinctive Features of the Cases

Chapter Three presented a comparative analysis, identifying broad patterns of excellence in others with regard to capabilities for operating in and through the IE, and it identified common practices and principles that contribute to that excellence. This chapter turns away from what can be learned from commonalities across the cases to explore what can be learned from the distinctive features of each case. See the companion volume for more detailed exposition and discussion of each case highlighted here.[1]

Israel

Israel's context and the principles by which the IDF operate both have several distinctive features. Israel and the IDF face a near-constant threat and so are nearly continuously operating both in the physical domains and in and through the IE. Though challenging, this creates a crucible in which principles and practice are tried, tested, and refined on a continual basis.

While Israel's traditional narrative has been a David-and-Goliath story, with plucky Israel fighting for survival against an angry Arab world that would like to see it pushed into the Mediterranean, recent decades have seen the narrative reversed, with plucky and disadvantaged Palestinians being oppressed and denied their rights by mighty Israel. This latter narrative has resonated not only in the Arab world but also in the West, leaving Israel feeling isolated and unfairly maligned. Israel has considered three potential audiences for its efforts to inform and influence: the Palestinians and the broader Arab and Islamic world, the West, and its own citizens. It has abandoned efforts to influence the first audience, perceiving that fundamental disagreements that buttress the opposition and contestation will not be resolved, improved, or deepened through Israeli efforts in the IE. Israel also perceives limited opportunity to persuade the second audience, assuming (rightly or wrongly) that Western audiences are biased against it and unlikely to offer further support. (Israel cites the boycott, divestment, and sanctions movement, which is popular in Western Europe, as evidence of this

[1] Paul et al., 2018.

claim.)[2] It recognizes, however, that it needs to avoid international sanctions or further opprobrium and so must refute propaganda or claims of Israeli atrocities, war crimes, or other excesses. The third audience, Israel's domestic audience, is of paramount importance. Keeping the support of its domestic audience, keeping its domestic audience informed, and keeping physical operations below the threshold of intensity or avoiding repression that might trigger a backlash and withdrawal of domestic support are Israel's priorities.

As a consequence of their perceived relationships with these three audiences, Israeli information and communication efforts have a distinct emphasis. They seldom seek to gain advantage through persuasion or influence in the IE; instead, they seek to minimize the impact of others' influence efforts on the audiences they care about and to minimize adverse consequences through the IE from their physical operations on the audiences that they care about. To be clear, Israel is very careful to conduct its operations in such a way that it does not alienate its domestic audience and to document those operations so that it can refute false claims that those operations crossed lines or violated international law—both of which might alienate its domestic constituency. Israel does not particularly care whether those physical operations alienate other audiences, however.

Key insights and lessons from Israel include the following:

- Operations must be planned with their consequences in and through the IE in mind.
- Efforts to communicate and inform must be coordinated across the entire government to avoid information fratricide, document and explain operations to domestic constituencies, and compellingly refute false claims that might undermine support domestically.
- Embedded press, combat camera or equivalent, independent journalists, and other sources of transparent reporting and verification are absolutely critical to insulate against falsehoods and propaganda.

NATO

NATO is distinctive because it is an alliance rather than a single national or organizational actor. It has no force structure of its own, and it is beset with the committee problem, with any aggregation of NATO forces beholden not only to NATO policies and procedures but also to the constraints imposed by the contributing nations. Because of these constraints, NATO restricts inform and influence efforts to strictly

---

[2] "Israel Bans Five U.S. Jewish, Christian and Muslim Leaders Backing Boycott Effort," *The Guardian*, July 25, 2017.

white information (discussed in Chapter Three). NATO aggregates all inform and influence capabilities under "strategic communications" as a coordinating and integrating body aimed at informing the domestic audiences of its 28 member nations and maintaining their support for NATO operations.

NATO's practical experiences from its extended commitments in Kosovo and Afghanistan have provided useful lessons for seeking effects in and through the IE under the restrictions imposed by the alliance and its member nations. Efforts in the IE in these two campaigns focused on gaining and maintaining the support of populations—both in the area of operations and in NATO member nations. Operations in these campaigns focused on truthful communication and were integrated with efforts to generate positive and favorable truthful information. So, civil-military operations would often be the primary effort, with other IRCs advertising and reinforcing good deeds and international goodwill and generosity.

Recent Russian aggression in Eastern Europe has sparked some interest in dormant NATO concepts. For much of the Cold War, NATO doctrine included the concept of "psychological defense," a capability area aimed at protecting troops and citizens from the propaganda and influence efforts of NATO's foes. While NATO never actually fielded meaningful capabilities under this concept, the concept itself remains potentially useful, and new joint capabilities have been under development through NATO's centers of excellence and other structures, such as in cyber defense.

Key insights and lessons from NATO include the following:

- The notion of psychological defense, though a capability area never fully developed by NATO, is both relevant and promising in the contemporary IE.
- The restriction to white information, the various constraints imposed by member states, and the fact that many members have very limited IRCs seriously constrain the potential effectiveness of NATO in and through the IE, but they provide significant institutional leverage when allies agree on a common approach.
- A commitment to white information makes it easier to integrate communication and information capabilities across capability areas, across branches of government, and across different nations' forces and governments, possibly leading to more successful engagement of local populations.

Canada

Canada's military is relatively modest in scale and scope. Historically, Canada's force structure and doctrine closely mirrored that of the United States, but it has moved away from that model. Current CAF concepts emphasize interoperability with NATO and Five Eyes (Australia, Canada, New Zealand, UK, and U.S.) forces, with CAF primarily acting as a supporting or enabling force rather than overly focused on constructing a narrative of Canadian force dominance. With that in mind, CAF IO and IRCs are also viewed as having a supporting and enabling role. Because of the military's modest size, there is no standing IRC force structure. Thus, these enablers are integrated on an as-needed and situational basis. When necessary, the CAF would mobilize its Influence Activities Task Force to fill training gaps and support staffing requirements to prepare deploying units for operations. This would have the virtue of activating several IRCs at once, with the attendant ease of deconfliction and synchronization.

Because it has no standing IRC force structure, Canada is trying to standardize and routinize its efforts in the IE through existing standing practices and processes, such as targeting. The logic is that even if expertise is not directly on hand, if the process requires consideration, then necessary expertise will be sought to complete the process.

Key insights from the Canadian case include the following:

- Bringing all IRCs together in a single task force dramatically eases the task of integrating and coordinating them.
- Routinizing and standardizing the role of IRCs in operations in the IE ensures that these capabilities receive consideration, despite the lack of a strong IE/IO/IRC advocate on the staff.
- IO is a joint function and requires coordination across multiple service platforms; the targeting process may serve as a useful mechanism for achieving strategic and operational objectives in the IE.

Germany

German operations in and through the IE are very robust by NATO standards but are highly constrained by the memory of Germany's past. Germans are well-educated about the strategies used by the Nazi and communist regimes to influence both domestic and foreign audiences and thus reject any kind of disinformation. As a result, Germany has some of strictest transparency and privacy laws in the world. So, while recognizing the importance of information and the IE, German government and Bundeswehr efforts are committed to honesty, transparency, and avoidance of even the appearance of manipulation or propaganda. Within those constraints, Germany is recognized for excellence in IO within NATO (trailing only the United States and United Kingdom), and it helps with training for allied IRC formations.

Since 2014, Germany's Center for Operational Communication has coordinated the country's information activities. The center focuses on working with local populations in areas of operations, assessing perceptions, and contributing to Bundeswehr communication campaigns. Germany's focus on local populations includes significant cultural and language training for troops deployed to various areas, as well as analysts working at the headquarters.

Germany is heavily engaged in social media and has a mature and robust social media outreach capability with a substantial following. It has also emphasized internal communication with its troops by establishing a dedicated radio channel and other platforms.

Key insights from Germany include the following:

- Complete rejection of falsehoods, half-truths, and propaganda makes it easy for Germany to integrate its communication activities and make significant use of social media, all without garnering suspicion or reproach. It also allows Germany to build a high degree of trust among its domestic constituencies.
- Germany has a history of engaging in psychological defense (protecting troops and citizens from the propaganda and influence efforts of foes), a concept that could be useful to the United States in the contemporary IE.
- An emphasis on cultural and language training is key to Germany's successes.
- Germany has been successful in enforcing a strict separation of efforts to address foreign and domestic audiences, whereby domestic messaging has purely informative value and may not be presented in a way to favor specific outcomes.
- German integration of IO-related capabilities does not include EW or cyber defense, which have distinct capabilities and resources.

China

China's efforts in the IE have more similarities to Russian efforts than might be expected. Both China and Russia emphasize deception and stratagem, and both employ a concept that involves conveying selected information to targets to get those targets to act in a way that the target perceives as being in its own interests but that actually serves China or Russia's interest.

China is unique in the extent of its investment in and emphasis on capabilities for action in the IE; it was the only case in our study with an entire branch of service–equivalent dedicated to these capabilities. China's efforts in the IE are virtually unconstrained. Its domestic audience is a major point of emphasis, and it is not required to respect this audience's privacy, nor is it barred from propagandizing, manipulating, or withholding information. China also engages in civil-military operations and civil affairs–like projects, especially in parts of Latin America and Africa, where its investments in infrastructure projects are intended to curry favor with government leaders and domestic populations.[3] The hope is that investment will provide entrée and create a positive impression of China and Chinese influence outside of East Asia.

China is broadly unconstrained in its use of psychological warfare outside its own borders, with no restrictions based on audience, the objective pursued, or whether recognized and active hostilities are taking place. China's "three warfares" (psychological warfare, public opinion warfare, and legal warfare) are distinctive in that none of the three would be recognized as warfare under traditional U.S. military thinking but, rather, as an aggressive form of peacetime competition.[4] Though, as Sun Tzu said, "To win without fighting is the acme of skill."

What is concerning about China's concepts for operations in the IE is that it has learned from the United States and cites the three warfares as an extension of observations of U.S. actions.[5] How we see ourselves is certainly not how others see us, and the lessons we learn from our experiences can differ dramatically from the lessons others draw from watching events unfold.

Key insights from China include the following:

- Being unconstrained by truth, audience, and declaration of hostilities makes China's range of military operations much broader than that of the United States and allows it to operate unopposed in the IE in many respects. While the United States cannot and should not operate free of all constraints in the IE, there are some restrictions that can and should be loosened. The United States must be prepared to contend with others that operate without these constraints.
- China's level of investment, degree of prioritization, and strategy of using information to drive targets toward courses of action that favor China are all worth emulating, even if its ethical standards, treatment of its domestic audience, and adherence to international norms are not.
- In the Chinese force structure, a "political work department" officer ranks above the regular staff and has direct access to the commander. In the organizational chart, the position is almost identical in placement to a PA officer on a U.S. staff. The political work department officer is one part commissar and one part full-spectrum information warfare adviser. This senior staff role helps Chinese forces ensure that information considerations are always part of planning and execution.

---

[3] For more on Chinese involvement in Africa, see Thrall, 2015, and Hanauer and Morris, 2014. For background on Chinese investments in Latin America, see Dollar, 2017.

[4] For further details, see Paul et al., 2018, Chapter Five.

[5] China's "three warfares" doctrine is separate, but related to, the Chinese concept of *total warfare,* which, itself, is a consequence of the related concept of *comprehensive national power*. For more, see Sean Golden, "China's Perception of Risk and the Concept of Comprehensive National Power," *Copenhagen Journal of Asian Studies*, Vol. 29, No. 2, 2011.

North Korea

North Korea's continued status as a hermit kingdom makes it highly distinctive, especially in a contemporary IE characterized by increasing connectivity and ubiquity. North Korea's considerable investment in information and population control has produced an almost fanatically committed and supportive domestic constituency. Most of the country's IRCs are dual-use—capabilities that it can use to monitor and control its population and turn toward other nations. North Korean OPSEC is unparalleled, as the threat of collective punishment or being sent to a labor camp helps reduce the human factor in security violations. While there is much that is distinctive about the country's capabilities and concepts for the use of the IE, there is little worth emulating, as what is effective is also antithetical to U.S. values.

Key insights and lessons from North Korea include the following:

- Because of the extent of information and population control, North Korean civilians are unreasonably committed to and supportive of their nation and its policies.
- Any effort by an external actor to influence audiences inside North Korea will face numerous challenges. However, the pressure of conflicting information from outside and the appeal of global popular culture can create wedges that, over time, might enable influence.

Iran

Iran is overmatched physically by its principal potential antagonists, driving it to seek asymmetric advantages. And it has done so: Iran engages heavily in information warfare, irregular warfare, and political warfare, and it is most distinctive in its use of international proxies. Iran supports Shia populations and nonstate irregular forces in Iraq, Yemen, Lebanon, Syria, Bahrain, and elsewhere; although these proxy forces are often only partially controlled by Iran, the threat and reality of these proxies have powerful effects in the IE. Iran has begun to make significant investments in capabilities for operations in cyberspace. Like other state actors, it uses its cyber capabilities in aggressive ways. Iran also uses its cyber capabilities more often for signaling and deterrence than for true aggression. By giving potential adversaries an idea of its capability, Iran hopes to dissuade certain kinds of aggression. Finally, Iran engages in civil-military operations and civil affairs–like activities through its patronage in the Middle East, but especially in Syria since the civil war began in 2011.

Key insights and lessons from Iran include the following:

- The Iranian Revolutionary Guards Corps Quds Force is an elite military unit used by the regime to train proxy forces abroad as a means of extending Iranian foreign and security policy throughout the Middle East.
- Deterrence involves both having punitive capabilities and demonstrating resolve and willingness to use them. Iran has used technical IRCs in exactly this way.

Russia

Engaging opponents in the IE is not part of a forgotten Cold War past in Russia; rather, it is a key component of its current foreign *and* domestic policy alike. The past few years have illustrated the breadth and depth of the Russian information apparatus, which has nimbly adapted to the digital age with significant investment in efforts to influence public opinion online and in foreign countries. The Russian Federation has adopted Soviet-era concepts, such as *maskirovka* and reflexive control, and found that they are even more effective in the contemporary IE than they were in the past.

Despite its economic woes, Russia has reinforced its military by developing rapid-reaction, highly deployable units alongside well-trained information warfare staff and a state-controlled media infrastructure, tight controls on information shared with the public, and a brutal strategy of intimidating and silencing its critics. These developments pose significant challenges to the United States and its allies—both in terms of avoiding a military confrontation on NATO's eastern flank and internally, given the significant influence capabilities that Russian actors and their proxies have built in many allied countries.

Key insights and lessons from Russia include the following:

- Russia has adopted a "firehose of falsehood" approach to propaganda, which emphasizes volume, rapidity, and frequency while eschewing any commitment to truth or consistency. Alarmingly, research in social psychology suggests that this approach can be highly effective.
- Russia's refusal to be bound by ethics, norms, or laws, along with heavy investment in IRCs, gives it incredible flexibility in the IE.
- Gray-zone aggression (aggression below the threshold of conflict but outside of routine competition) must be countered in the gray zone and fought at least partially in the IE. Unfortunately, current Army (and broader DoD) authorities hinge too much on a false dichotomy between peace and war, preventing the Army from employing many of its IRCs to counter Russian gray-zone aggression.

Hezbollah

Of course, nonstate actors are distinct from state actors, but like state actors, nonstate actors will copy the successful blueprints of others. Hezbollah has the distinction of being the originator of the "information-first" operational mindset. Its mantra has *for decades* been "If you haven't captured it on film, you haven't fought." Its general operational principles and principles for operations in and through the IE are one and the same. From Hezbollah's perspective, physical combat power and social service provision are just additional IRCs. All Hezbollah members and operators are instructed in the importance of effects in and through the IE, and it invests heavily (for a nonstate actor) in IRCs, particularly different channels and media for communication and propaganda dissemination. Also distinctive is its responsive opportunism. During the 2006 conflict with Israel, within minutes of Hezbollah hitting the Israeli naval destroyer *Hanit* with a missile, the group's secretary general, Hassan Nasrallah, announced the strike on al-Manar with accompanying footage for distribution by regional media and YouTube. It took Israel hours to provide an official response.

Key insights and lessons from Hezbollah include the following:

- The emphasis on information is embedded in planning at all levels and inculcated in the culture of the military arm of Hezbollah.
- More than any other nonstate actor, Hezbollah appreciates the importance of efforts in and through the IE and treats them as a warfighting function.
- Responsive opportunism: Hezbollah is always ready to take advantage of any operating context and the different actors therein.

Al-Qaeda

Al-Qaeda is infamous and distinctive for its intuitive understanding of the incalculably large impact in the IE of spectacular physical attacks. Since the peak of its power, the group has ceded ground in many respects (including in the IE) to newer and more sophisticated nonstate actors. Despite its decline and reduced ability to produce propaganda, the sophistication and appeal of al-Qaeda's propaganda help it remain relevant, archived in perpetuity on the Internet.

Key insights and lessons from al-Qaeda include the following:

- The group has an intuitive understanding of the information value of kinetic operations and the fact that kinetic actions can have orders of magnitude greater impact on the IE.
- Deterring or destroying a physical organization does not necessarily put an end to its influence, as older but high-quality propaganda preserved online can continue to circulate for years.

ISIL/Daesh

ISIL is distinctive as a nonstate actor with aspirations of statehood, and those aspirations have implications for the IE. Its central narrative of the caliphate reborn is incredibly compelling, especially while the group appeared to have successfully established a quasi-state to which it could apply that name. ISIL maintains a high level of message discipline, but it does not need to work too hard to accomplish that, because its themes are few, focused, and compelling. The organization also emphasizes the importance of information: Its leadership invests in media capabilities that have high production values and acknowledges the importance of media personnel by according them the rank of "emir." These media units combine the group's information power capabilities, including its inform and influence capabilities and its cyber capabilities.

Key insights and lessons from ISIL include the following:

- The caliphate narrative is incredibly effective, both for unifying ISIL operations and messages and for providing a compelling frame for those operations for supporters and potential supporters.
- Cleanly grouped and tightly focused themes make message discipline easy.

Mexican Drug-Trafficking Organizations

Most distinctive among the nonstate actors of concern in our study and of an entirely different ilk are Mexico's DTOs. These organizations generally lack a formal capability structure for operations in the IE. Nonetheless, in narrow and specific ways, they are highly effective there. This is because of the easily understood connection between DTO violence and the goal of intimidation and coercion: Every act of violence is also a message, with perfect consistency. Violence is both the medium and the message. The imperative to protect an organization's reputation and to maintain its reputation as inescapable, unavoidable, and undefeatable follows the logic of deterrence, an effect in the IE that is familiar to all of the state actors in our case studies.

The DTOs are also distinctive in their intense recruitment and indoctrination efforts, with rites and an intensity resembling violent extremist organizations or Western militaries, in which civilians are torn down in boot camp and rebuilt as soldiers.

Although the obscene violence of DTOs makes it difficult to draw positive lessons or insights that might be of use to the U.S. Army, there is something to which the Army might aspire. Could the Army achieve the same level of integration, in which maneuver and fires send messages that are perfectly aligned with operational goals? For the Army, violence is not both the medium and the message, but could the physical and informational objectives of Army operations be so tightly woven that soldiers pursuing one will unavoidably pursue the other? If so, it would certainly make success in and through the IE easier to achieve.

CHAPTER FIVE

Conclusions and Recommendations

This report provided an overview of the principles and practices of others to make observations and draw lessons that will potentially be useful to future U.S. Army efforts in and through the IE. We reviewed the IE organization and efforts of 12 others: four allied nations, four state actors of concern, and four nonstate actors of concern. That review revealed a number of principles and practices generally associated with effectiveness in and through the IE. Specifically, as described in Chapter Three:

- generous resourcing of information power and IRCs
- giving prominence to information effects in operational planning and execution
- increasing the prestige and regard of IRC personnel
- integrating physical and information power
- extensively using information power in operations below the threshold of conflict
- employing the concept of getting the target to unwittingly choose one's preferred course of action
- carefully recording and documenting one's own operations, and enabling embedded press, independent journalists, and combat camera (or an equivalent) as witnesses in the battlespace
- high production values.

Our 12 case studies also revealed some noteworthy observations that are unique to specific cases. As detailed in Chapter Four, these include the following:

- Benefits accrue when efforts in and through the IE are coordinated across the equivalent of agencies and departments.
- The concept of "psychological defense" of soldiers and citizens is both relevant and promising in the contemporary operating environment.
- Benefits accrue when IRCs are physically and organizationally colocated.
- Benefits accrue when the processes and capabilities for delivering effects in and through the IE are part of standard military routines and processes.
- Deterrence involves both having capabilities and demonstrating resolve; aggressive employment of technical IRCs can demonstrate both capability and resolve.

- The principle of "initiative" is available in the IE but requires a certain amount of opportunism.
- Kinetic action, especially spectacular kinetic action, has incredible information power, too.
- High-quality and compelling propaganda can live forever on the Internet.
- Narrative can be incredibly powerful.

Taken together and considered in light of current U.S. Army capabilities for operations in and through the IE, these findings have several implications. First, Army should adopt as many of the effective practices as possible, eschewing only those that are inconsistent with U.S. laws or values.

Second, there are gaps between Army capabilities and the best-in-breed capabilities observed in our case studies. Simply put, the Army is not currently the best at everything. This is not a surprise, and it was part of the motivation for this study. In seeking to close these gaps, however, it does beg an interesting question: Does the Army need to be the best at everything, or can some gaps be accepted, effectively ceding dominance in some capability areas to actors of concern? While this is obviously not ideal, it may be necessary due to fiscal constraints. The Army should consider conducting cost analyses of the requirements associated with closing or eliminating these gaps, as well analyses of the risks associated with leaving some gaps open.

Third, where there are gaps between Army capabilities and best-in-breed capabilities observed among other actors, there appear to be three reasons.

- Some of the capability gaps are actually *capacity* gaps in that technical capability or capability proficiency is similar but others have invested more and thus have a great deal more of the relevant capability.
- Some of the capability gaps are actually *conceptual* gaps, in which others have adopted effective principles and practices that the Army has not yet adopted but could.
- Some gaps are due to differences in constraints. Many others operate in a virtually unconstrained fashion, while the Army operates constrained by authority, statute, policy, and ethics.

Obviously, the first two reasons for capability gaps are more amenable to gap closure; gaps stemming from the illegal or unethical practices of others must be surmounted in other ways.

Recommendations

Based on these findings, our holistic understanding of the cases (including the U.S. Army baseline case), the relative merits of the different cases and the practices associated with them (as described in Chapter Three), and our years of experience analyzing IO and IRCs in Army, joint, and international contexts, we make the following recommendations.

Because many of the gaps between Army capabilities and the best-in-breed capabilities of others are conceptual, the Army should adopt and emphasize new concepts. Since successful actors give prominence to information effects in manning, planning, and execution, the Army should *give effects in and through the IE more staffing and priority.* Such effects should play a greater role in, and receive more attention during, the planning and execution of future plans and operations. To enable this needed shift, changes will be required in doctrine, training, and education to emphasize the role of the IE in determining the outcomes of operations, as well as the importance of planning effects in and through the IE to achieve desired outcomes and end states.

Similarly, since highly effective actors operate under concepts that integrate physical and informational power, the Army should change its culture, doctrine, training, education, and planning processes to *promote a view of information power as part of combined arms.* Physical combat power and informational combat power should go hand in hand in all aspects of Army operations. This will require a significant change in soldiers' mindset, including recognition that kinetic actions affect the IE while action in the IE drives the motivation for adversary kinetic action, as well as the recognition that the full spectrum of effects can be leveraged to affect adversaries and their supporters (in the form of physical destruction, attrition, demonstration of force, command-and-control warfare, and shaping legitimacy and support from relevant constituents).

The IO concept as currently employed by Army and joint forces aspires to integrate IRCs in support of operations. Unfortunately, as currently executed, IO usually fall well short of the full integration of physical and informational combat power achieved by effective actors and captured by our recommendation information power should be part of combined arms. Too often, information-related concerns follow their own staffing processes and are removed from the core work of the planning and operations staff, with the IE an ancillary consideration and IRCs sprinkled on at the last minute or treated as an underappreciated bolt-on to fires. Effective actors include effects in the IE as part of the business of their core staff. The full embrace of effects in and through the IE as part of combined arms will necessitate that the Army *routinize and standardize the processes associated with operations in and through the IE to be consistent with and part of other routine staff processes,* including, targeting, planning, assessment, intelligence, and operations.

In coordination with embracing information power as part of combined arms and creating situations in which the mission is the message, *Army commanders and staffs should make it easier for soldiers' actions to support desired effects in the IE by tying political, physical, and cognitive objectives together coherently in plans and clearly communicating those compound objectives to maneuver forces.*

Because effective actors make extensive use of information power in their operations below the threshold of conflict, and because current DoD and Army authorities for such operations are heavily constrained, we recommend expanding such authorities. In coordination with DoD and the U.S. government, the Army should *seek expanded authorities to operate in the IE short of declared hostilities*—that is, gray-zone conflicts and steady-state/phase 0 operations. Furthermore, if effects in the IE are important in steady-state/phase 0 (and they are), and if Army is given increased latitude to operate in phase 0, then it will need to *bring more IO, MISO, and other IRC capability out of the reserves.*[1]

To better support integration, in coordination with DoD, the Army should *tear down or move the firewall between PA and IO and other IRCs.*[2] This change should not allow PA to misrepresent, but it also should not continue to excuse it from integrating and coordinating with influence efforts and supporting overall operational objectives.

Quantity has a quality all its own, and effective actors invest much more generously in information power than the U.S. Army currently does as a proportion of their overall spending, in absolute terms, or both. To close gaps with others, the Army needs greater IO and IRC capacity. We recognize that fiscal constraints make it difficult to increase force structure, but we nonetheless recommend that the Army *take steps to close capacity gaps in key capability areas, including cyber, influence, OPSEC, and MILDEC.* Because effective actors give IRC personnel high regard in their organizations, steps to increase Army IRC capability should also include changes to *make IO and IRC career fields and military occupation specialties larger, more attractive, and more prestigious.* The Army should also consider creating more IE-related branches—not just functional areas—and expanding the use of warrant officers in IO and the IRCs.

---

[1] Currently, a significant fraction of the IO and MISO force is in the reserves. Active-duty MISO forces support special operations, while reserve MISO formations support general-purpose forces. If DoD were to participate more robustly in the IE short of declared hostilities, the reserve component would quickly be exhausted by demands for support to such steady-state operations.

[2] *Firewall* is the term often used to describe the artificial divide between PA and other IRCs. Since all military communicators are prohibited from misleading the U.S. public, and the primary audience for these efforts is the U.S. public, some PA officers interpret this as an excuse to avoid meeting with or having any awareness of the other IRCs. This situation dramatically increases the prospects for information fratricide, where different capabilities transmit conflicting information.

References

Cheng, Dean, "Winning Without Fighting: The Chinese Psychological Warfare Challenge," Washington, D.C.: Heritage Foundation, Backgrounder No. 2821, July 11, 2013. As of January 30, 2018:
http://www.heritage.org/global-politics/report/winning-without-fighting-the-chinese-psychological-warfare-challenge

Clarke, Colin P., and Chad C. Serena, "To Defeat ISIL's Brand, Its Territory Must Be Reclaimed," *National Interest*, July 8, 2016. As of January 30, 2018:
http://nationalinterest.org/blog/the-buzz/defeat-isils-brand-its-territory-must-be-reclaimed-16890

Dollar, David, *China's Investment in Latin America*, Washington, D.C.: Brookings Institution, Geoeconomics and Global Issues Paper 4, January 2017. As of January 30, 2018:
https://www.brookings.edu/wp-content/uploads/2017/01/fp_201701_china_investment_lat_am.pdf

Gartenstein-Ross, Daveed, Nathaniel Barr, and Bridget Moreng, "How the Islamic State's Propaganda Feeds into Its Global Expansion Efforts," *War on the Rocks,* April 28, 2016. As of January 30, 2018:
https://warontherocks.com/2016/04/how-islamic-states-propaganda-feeds-into-its-global-expansion-efforts

Golden, Sean, "China's Perception of Risk and the Concept of Comprehensive National Power," *Copenhagen Journal of Asian Studies*, Vol. 29, No. 2, 2011, pp. 79–109.

Hanauer, Larry, and Lyle J. Morris, *Chinese Engagement in Africa: Drivers, Reactions, and Implications for U.S. Policy*, Santa Monica, Calif.: RAND Corporation, RR-521-OSD, 2014. As of January 30, 2018:
https://www.rand.org/pubs/research_reports/RR521.html

"Israel Bans Five U.S. Jewish, Christian and Muslim Leaders Backing Boycott Effort," *The Guardian*, July 25, 2017. As of January 30, 2018:
https://www.theguardian.com/world/2017/jul/25/israel-boycott-bds-movement-law-leaders-barred

Kramer, Andrew E., "More of Kremlin's Opponents Are Ending Up Dead," *New York Times,* August 20, 2016.

Majidyar, Ahmad, "Celebrations of Iranian Revolution Across Syria Shows [sic] Iran's Soft Power Hegemony," Washington, D.C.: Middle East Institute, February 13, 2017. As of January 30, 2018:
http://www.mideasti.org/content/io/celebrations-iranian-revolution-across-syria-shows-iran-s-soft-power-hegemony

Mazarr, Michael J., *Mastering the Gray Zone: Understanding a Changing Era of Conflict*, Carlisle Barracks, Pa.: U.S. Army War College, Strategic Studies Institute, 2015.

Miller, Greg, and Souad Mekhennet, "Inside the Surreal World of the Islamic State's Propaganda Machine," *Washington Post*, November 20, 2015.

Munoz, Arturo, and Erin Dick, *Information Operations: The Imperative of Doctrine Harmonization and Measures of Effectiveness*, Santa Monica, Calif.: RAND Corporation, PE-128-OSD, 2015. As of January 30, 2018:
https://www.rand.org/pubs/perspectives/PE128.html

Nicholas, James, "Australia: Current Developments in Australian Army Information Operations," *IO Sphere*, Special Edition 2008, pp. 38–43.

Paul, Christopher, *Information Operations Doctrine and Practice: A Reference Handbook*, Westport, Conn.: Praeger Security International, 2008.

———, *Strategic Communication: Origins, Concepts, and Current Debates*, Santa Barbara, Calif.: Praeger, 2011.

Paul, Christopher, Colin P. Clarke, Michael Schwille, Jakub P. Hlavka, Michael A. Brown, Steven S. Davenport, Isaac R. Porche III, and Joel Harding, *Lessons from Others for Future U.S. Army Operations in and Through the Information Environment: Case Studies,* Santa Monica, Calif.: RAND Corporation, RR-1925/2-A, 2018. As of April 2018:
https://www.rand.org/pubs/research_reports/RR1925z2.html

Paul, Christopher, Kristen Sproat Colley, and Laura Steckman, "Fighting Against, with, and Through Narrative," *Marine Corps Gazette*, forthcoming.

Paul, Christopher, and Miriam Matthews, *The Russian "Firehose of Falsehood" Propaganda Model: Why It Might Work and Options to Counter It,* Santa Monica, Calif.: RAND Corporation, PE-198-OSD, 2016. As of January 30, 2018:
http://www.rand.org/pubs/perspectives/PE198.html

Pomerantsev, Peter, "Russia and the Menace of Unreality: How Vladimir Putin Is Revolutionizing Information Warfare," *The Atlantic*, September 9, 2014. As of January 30, 2018:
http://www.theatlantic.com/international/archive/2014/09/
russia-putin-revolutionizing-information-warfare/379880

Porche, Isaac R. III, Christopher Paul, Michael York, Chad C. Serena, Jerry M. Sollinger, Elliot Axelband, Endy M. Daehner, and Bruce J. Held, *Redefining Information Warfare Boundaries for an Army in a Wireless World*, Santa Monica, Calif.: RAND Corporation, MG-1113-A, 2013. As of January 30, 2018:
http://www.rand.org/pubs/monographs/MG1113.html

Porche, Isaac R. III, Bradley Wilson, Erin-Elizabeth Johnson, Shane Tierney, and Evan Saltzman, *Data Flood: Helping the Navy Address the Rising Tide of Sensor Information*, Santa Monica, Calif.: RAND Corporation, RR-315-NAVY, 2014. As of January 30, 2018:
https://www.rand.org/pubs/research_reports/RR315.html

"Propaganda," *Merriam-Webster Online Dictionary*, undated. As of January 30, 2018:
https://www.merriam-webster.com/dictionary/propaganda

Reporters Without Borders, "China," web page, undated. As of January 30, 2018:
https://rsf.org/en/china?nl=ok

Seftel, Bennett, "What Drives ISIS," *Cipher Brief*, May 5, 2016. As of January 30, 2018:
https://www.thecipherbrief.com/article/middle-east/what-drives-isis-1089

Siers, Rhea, "Israel's Cyber Capabilities," *Cipher Brief*, December 28, 2015. As of January 30, 2018:
https://www.thecipherbrief.com/article/israel's-cyber-capabilities

Thomas, Timothy L., "Asia-Pacific: China's Concept of Military Strategy," *Parameters*, Vol. 44, No. 4, Winter 2014–2015, pp. 39–48.

———, "Russia's Military Strategy and Ukraine: Indirect, Asymmetric—and Putin-Led," *Journal of Slavic Military Studies*, Vol. 28, 2015, pp. 445–461.

Thrall, Lloyd, *China's Expanding African Relations: Implications for U.S. National Security*, Santa Monica, Calif.: RAND Corporation, RR-905-A, 2015. As of January 30, 2018: https://www.rand.org/pubs/research_reports/RR905.html

U.S. Department of Defense, *Department of Defense Strategy for Operations in the Information Environment*, Washington, D.C., June 2016. As of January 30, 2018: https://www.defense.gov/Portals/1/Documents/pubs/DoD-Strategy-for-Operations-in-the-IE-Signed-20160613.pdf

U.S. Joint Chiefs of Staff, *Information Operations*, Washington, D.C., Joint Publication 3-13, incorporating change 1, November 2014. As of January 30, 2018: http://www.jcs.mil/Portals/36/Documents/Doctrine/pubs/jp3_13.pdf

Watzlawick, Paul, Janet Beavin-Bavelas, and Don D. Jackson, *Pragmatics of Human Communication: A Study of Interactional Patterns, Pathologies and Paradoxes*, New York: W. W. Norton and Company, 1967.

Zelin, Aaron Y., "Picture or It Didn't Happen: A Snapshot of the Islamic State's Official Media Output," *Perspectives on Terrorism*, Vol. 9, No. 4, 2015.